序

近年，加入飯聚集團當煮人的人數與日俱增，因為他們相信，只要能成為一名出色的煮人，煮得一手好菜，就能俘虜他們的愛人以及一眾飯人。但要成為一名傑出煮人，得到飯人們的讚賞及支持，就得擁有永不失敗，保證成功的食譜。

江湖上一直流傳着，有一份無懈可擊的零失敗食譜，只要得到這些食譜，跟着食譜內的步驟及秘訣，就能煮出比「黯然銷魂飯」更好食及令人動容的菜式。

作為飯聚份子的召集人，為了令各煮人都得到幸福及快樂，我走遍了大小街市，尋覓過各種不同的食材，請教過各路烹飪達人，並重複又重複地，將這些零失敗食譜反覆鑽研及實習，終於研發出這一系列的零失敗食譜，並將它們集結及輯錄起來，與大家分享。

烹飪是愛的傳遞及分享，希望《零失敗的煮人》能在你家的餐桌派上用場，及能得到你們的囍愛，本人將感到十分榮幸。

街市
給我無限的烹飪靈感

無論出外旅遊或公幹，總會心思思往當地的街市走走。
每個街市都有不同的特色，既反映當地民風之餘，也反映了生活質素，同時也令我大開眼界，猶如走入大觀園。

香港街市是我的尋寶勝地，五花八門、琳琅滿目的食材、調味料、配料，往往讓我的烹飪靈感邊走邊發酵，菜式在腦中盤桓，走一圈，可能已構思了一圍檯的菜式了。

另一個讓我囍歡逛街市的原因，是人情味。
幫趁多年的舖頭，員工老闆已成為我的朋友，大家一起分享煮食心得，談天說地，走進街市，猶如赴老朋友的約會。

大家多逛街市吧！

目錄

Contents

Roast chicken stuffed with Kimchi
and Korean rice cake...200

Steamed grouper with black garlic, ham
and dates...200

Steamed pork ribs
in chilli black bean sauce...201

Homemade crispy roast pork belly...202

Ma La salted chicken salad with cucumber,
coriander and shrimp roe...203

Bone-softened Pacific saury
in tomato sauce...204

Braised pork belly with dried cauliflower
and porcini mushrooms...205

Crispy chicken in black bean
and olive honey glaze...206

Crispy roast chicken stuffed
with preserved clams and sand ginger...207

Home-style braised beef shin
in tomato sauce...208

Steamed pork ribs in Renren sauce...209

Noodles • Bread

Black garlic bread...210

Black truffle shrimp toast...211

Kimchi and semi-dried oyster pizza...212

Seafood angel hair pasta in Tom Yum sauce...214

Egg noodles dressed in Renren sauce,
garlic scallion oil and shrimp roe...215

Tom Yum seafood pizza...216

Clams in spicy wine sauce
with instant noodles...218

Soup

Double-steamed squab soup with black garlic,
peanuts and black-eyed beans...209

Black truffle soup with mushrooms...219

Tom Yum Goong pumpkin seafood soup...220

Wine • Desserts

Warm black truffle chocolate fondant
a la mode...222

Okinawa Kurozatou New Year Cake...222

Homemade Chinese plum wine...223

零失敗 獨門飯煲焗黑蒜

Zero-failure black garlic in rice cooker...174

健康食品的熱潮在香港像季候風一樣，不停地吹着，近年大熱的非黑蒜莫屬了，據說它含有的維生素比大蒜多，其氨基酸含量比白蒜多 5 倍、胡蘿蔔素比白蒜多 1.5 倍、抗氧化能力更是白蒜的 10 倍。

據研究說，黑蒜有益氣活血、促進新陳代謝、增強腸胃功能、調理血液循環、提高免疫力、促進身體血糖代謝分解、降血糖、消除體味體臭及防癌等食療功效。

做好的黑蒜可以用來燉湯、炒菜、蒸菜、伴沙律、混合橄欖油黑醋打成醬汁，甚至塗抹在烘麵包上等等都非常合適，食法多變而且美味，難怪現在成為健康飲食人士的新寵兒。

在市場上的現成黑蒜也不便宜，兩粒裝的黑蒜大約也要港幣 30 多元。但是，若在家裏自己製作就便宜得多了，而且並沒有甚麼難度，只要跟着以下步驟，我保證大家一定成功。何況，健康食品要長期食用，自家製作長食長有又方便。

材料

- 整粒連皮蒜頭或獨子蒜

工具

- 電飯煲
- 厚身疏孔的竹墊
  約 0.5cm 高，可在家品店買一般茶
  壺用的竹墊最好
- 小型竹篩

做法

1. 先在電飯煲最底部放入厚身的竹墊，在上面平均放上一層蒜頭，在蒜頭上面再放上一張竹篩，然後在竹篩上再平均放上一層蒜頭，如此重複至放滿電飯煲的內膽為止，蓋上電飯煲蓋，開着保溫掣，不要開蓋，讓蒜頭在電飯煲內 60 度左右的溫度內發酵 14 天至 20 天即成。

2. 經發酵後的蒜頭，剝開衣後蒜肉會變成全黑色（若仍呈栗子色的話，即代表未發酵完成），而且質地變軟，原來辛辣的生蒜味亦會轉化成甜味。將蒜頭取出，放在陰涼地方吹乾一至兩天，然後便可以放入密封及存氣的保鮮袋或容器保存，慢慢享用了！

貼士

1. 做黑蒜的蒜頭不要用水洗，及在發酵過程中切記不要打開煲蓋，以免影響發酵效果；而且電飯煲最好放在通風的地方。

2. 選蒜頭要揀選飽滿未發芽的，我喜愛揀選外皮帶紫色的品種，蒜味特別香濃，另外我亦很喜歡用獨子蒜，因為蒜肉飽滿大粒，而且味道較甜。因此，在每一層蒜頭之間較小的空位內，最好就放些獨子蒜以填滿這些空間。

3. 電飯煲最好選煲蓋可以鎖住的那種，以免中途不慎打開。

4. 在電飯煲最底層放上較厚的墊子，是因為發酵期長達兩至三星期，以免蒜頭直接接觸電飯煲較高溫的煲底而變焦。另外每層擺放的蒜頭亦要平均，盡量避免重疊，讓煲內溫度平均滲透，讓每顆蒜頭都可發酵至最佳效果。

5. 由於不同季節的氣候及潮濕度影響，有時候如做出來的黑蒜太濕的話，可將新鮮蒜頭先放入風乾機風乾兩小時或略為曬過才放入電飯煲，這樣做出來的黑蒜便會乾身一些。

椰菜花乾
~ 自家曬製的天然美味

Home-style sun-dried food bursting with natural flavours dried cauliflower...178

在我祖母還在的童年時代，家裏很多保存性食物都是自家製作的，如菜乾、臘肉、臘腸、葛粉、馬蹄粉之類都是她老人家的拿手之作。每年不同季節，祖母都有她不同的「產品」，但那時候自己還未懂得欣賞這些寶貴的飲食文化，直到長大後才發現，這些家庭手作仔真的又好玩又好吃。最重要的是這些食物還是全人手製作，純天然，無添加，健康，安全！

如天氣寒冷乾燥，是自家製作風乾蔬菜的最好時機，透過自然曬製而成的蔬菜，保存了蔬菜的原味，還添加了陽光的樸實清新風味，加進不同的菜式中，無論燜、煮、煲湯都別具特色風味。

香港常見的菜乾、醃菜包括梅菜、冬菜、白菜乾、蘿蔔乾（菜脯）、番薯乾等等。其實曬製這些菜乾十分簡單，像廣東人常用來煲老火湯的白菜乾，只要將新鮮白菜洗淨，用滾水煮一下，然後串起來在陽光下曝曬約一星期直至完全乾透即成。

與大家分享的天然生曬椰菜花乾，做法也十分簡單，若大家有興趣的話，我很建議大家嘗試一做，領略一下這種天然的風味。

材料

- 新鮮椰菜花 ………… 10 斤
- 粗鹽 ………… 約 3 湯匙

做法

1. 先將椰花切成一朵朵，煲滾水，下粗鹽煮溶，分次將椰菜花焓約 10 數秒，撈起，瀝乾水分。
2. 然後平均攤在透氣的筲箕上，在陽光下曝曬約一星期，直至椰菜花乾透為止，便可放入膠袋或器皿中，放入雪櫃保存，慢慢享用了。

貼士

1. 切椰菜花時不要切得太細，因為曬乾後的椰菜花體積會縮小，若切得太小的話，完成品便會變得更小「唔見食」了。
2. 椰菜花乾的煮法亦相當簡單，只要將椰菜花乾略為清洗，再用水略泡至膨脹及軟身，然後配上排骨、肉類煮湯、紅燒燜肉、咖喱、炒肉絲、煲粥等等都非常美味。

註：這食譜曾在《返屋企●食自己》刊登。下一個要分享的零失敗菜式「椰菜花乾牛肝菌燜腩肉」亦有採用椰菜花乾，為免讀者花時間撲食譜，故將這食譜再刊登。

豉香青辣醬

辣椒是茄科辣椒屬草本食物的果實，品種繁多，據講現在世界上的辣椒種類達千多種之多，它無論新鮮、曬乾或做成各類辣椒製品都用途多樣及廣泛，做泡菜、涼拌、炒菜、煲湯都少不了它，其鮮艷奪目的顏色，在伴碟裝飾上都佔有了相當重要的角色。青辣椒其實是辣椒果實未成熟即將變成紅辣椒前摘下來的青色未熟辣椒，它沒有紅辣椒那麼辛辣，一般來說顏色較深的會辣些，購買時應揀選色澤光亮而辣椒身觸感較結實的為佳。

這款以青辣椒做成的辣椒醬，口味清新獨特，辣度適中，它有豆豉的鹹香，亦有花椒油帶來的麻香川蜀風味，非常惹味醒胃，無論配以各式粥粉麵飯，做涼拌或家常小菜都非常合適！

材料

- 中型長身青辣椒 .... 1 公斤
（今次示範的是選用湖南青辣椒）
- 乾葱頭 .... 150 克
- 豆豉 .... 50 克
- 菜油 .... 300 毫升
- 花椒油 .... 50 毫升

調味料

- 砂糖 .... 4 湯匙
- 白醋 .... 3 湯匙

做法

1. 辣椒洗淨，抹乾水分，預熱焗爐約攝氏 250 度，將辣椒烘至微焦及表皮皺起，待涼卻後小心將辣椒皮剝去，剁碎；乾葱洗淨，剝去外皮，剁碎；豆豉洗淨，吸乾水分，剁碎備用。

2. 燒熱菜油，將乾葱茸炸炒至微微焦香，下豆豉炒勻，再下青辣椒碎炒勻，最後下花椒油及調味料拌勻即可關火，趁熱裝入已經消毒處理的容器，並立即將容器倒轉以達真空處理效果。

貼士

1. 焗爐一定要先預熱，務求青辣椒一放入焗爐，高溫便可直接燒到辣椒表皮，令辣椒皮盡快燒焦起皺，若青辣椒在焗爐內燒焗太長時間的話，青辣椒便會變得太熟，口感也會太腍，色澤亦不鮮艷，大大影響了辣椒醬成品的效果。

2. 此青辣椒醬更可依個人口味以沙爹醬或魚露代替豆豉，各有風味。另外，我亦嘗試過加入適量蝦米碎及瑤柱絲同炒，亦另有一番滋味，大家不妨試試。

3. 盡量揀選較豐厚的青辣椒，因為燒焗完之後的辣椒部分水分會揮發，肉太薄的話，青辣椒肉便「不見使」了，同時亦可將部分辣椒籽刮去，因為若太多辣椒籽在醬內的話，會影響口感。

私房秘製仁稔醬
Homemade Renren sauce...176

仁稔是非常傳統的廣東食品，用豉油、砂糖醃製好的仁稔，可以當成涼菜、涼果般直接食用，亦可取其醃製好的仁稔肉加些麵豉醬、甜酸子薑，用來蒸魚、蒸排骨以及蒸雞等等，非常惹味開胃。而用豉油浸泡過仁稔的仁稔汁，因為吸收了仁稔的奇特酸澀味道，亦是一道非常奇味的醬汁，你可以加些橄欖油或麻油調製成沙律汁，拌以各式沙律、涼拌前菜，或加些切碎的辣椒絲用來做沾醬用，更是美味非常；燜煮肉類，加些仁稔汁來調味，亦會令菜式增添妙不可言的風味。

仁稔除了用來醃製之外，亦可將新鮮仁稔肉加入子薑、切碎的五花腩、麵豉醬、辣椒等炒成仁稔醬，這亦是一道相當經典的廣東家常醬料，非常美味，可惜現在已極少人懂得做了，加上仁稔只在每年初夏約六月短短一個月左右出產，所以更見珍貴。分享此食譜，亦希望仁稔這款獨特而美味的傳統食品可以延續下去，讓更多喜愛烹飪的朋友能得以品嘗及欣賞！

舌尖上獨特的味覺體驗、獨一無二的酸、甜、鹹、辣、香風味，讓你回到童年時代純真簡樸、閒話家常的飯桌上，無數的味道回憶，彷彿仍在我的舌頭上徘徊！

材料

- 仁稔 .... 4 斤半
- 白醋 .... 200 毫升
  醃仁稔用
- 子薑 .... 3 斤
- 粗鹽 .... 60 克
  醃子薑用
- 五花腩 .... 2 斤半
  連皮
- 蝦米 .... 320 克
- 片糖或黃砂糖 ...... 620 克
  片糖切碎
- 蒜頭 .... 約 10 粒
  切片
- 指天椒 .... 50 克
  切圈，約 20 隻，
  視乎愛辣程度而增多或減少
- 白米醋 .... 200 克
- 廖孖記麵豉醬 ...... 850 克
- 鹽 .... 2 茶匙
  調味用
- 紹興酒 .... 1/4 杯
- 浸蝦米水 .... 1 杯
- 油 .... 300 毫升

做法

1. 仁稔洗淨，切去蒂，十字剝開，起出仁稔肉，倒入白醋撈勻，醃約 1 小時，隔起白醋，備用。

2. 子薑洗淨，切成細粒，加入粗鹽 60 克撈勻，醃約 1 小時，讓它出水，然後用手將子薑的水分揸出，備用。

3. 五花腩洗淨，如豬皮上還有毛的話，要用廚用火槍燒掉，然後連皮蒸約 50 分鐘至熟，待涼卻後連皮切成細粒，備用。

4. 蝦米洗淨，用水浸約 1 小時至軟身，隔起，瀝乾水分，略為切碎，留起約一杯份量的浸蝦米水，備用。

5. 乾鑊燒熱，將已揸乾水分的子薑粒炒至水分蒸發，盛起備用。

6. 抹淨鑊後，加入仁稔，同樣以乾鑊將仁稔粒炒至變褐黃色，盛起備用。

7. 抹淨鑊，下油 300 毫升，先下蒜片及辣椒爆香，再下蝦米炒至散出香氣及油開始起泡，盛起，備用。

8. 另備一較大的鍋燒熱，加入豬肉炒熱至開始滲出油分，倒入紹興酒炒勻，加入麵豉醬及片糖碎炒勻，至片糖溶化，然後加入仁稔、子薑、蝦米、指天椒、蒜片、鹽 2 茶匙及蝦米水，不停炒動約 10 分鐘，至仁稔醬開始濃稠及色澤變得光亮「令」身，即可關火，盛起涼卻後便可裝入密封容器，放入雪櫃保存了。

貼士

1. 選用片糖做出來的仁稔醬色澤會深些，及帶有一股焦香味道。而選用黃砂糖做出來的仁稔醬會濃稠些及色澤會光亮一些。

2. 因仁稔醬較為濃稠及較多糖分，所以炒製時要不停炒動，以避免黏鍋及變焦。

3. 當仁稔醬炒至色澤變得光亮「令」身時，便代表仁稔醬已經完成，此時你便可以關火；如個人喜愛仁稔醬較濃稠「傑」身一些的話，炒的時間再長一點便可以了。

養生黑蒜
海鮮田園沙律
Seafood salad with black garlic...179

其實，黑蒜在用於中西菜式上的可塑性都相當高，一般家庭菜可作蒸雞、蒸排骨等，在小炒菜式臨完成前加入幾粒黑蒜快炒後才上碟，馬上增添了一種養生的健康味道。如想再簡單一些，這道健怡沙律也是一個很好的選擇，好味！健康！快、靚、正！

材料

- 黑蒜..............約兩整粒

  去衣取蒜肉，一開四，留適量用叉壓爛成黑蒜茸作沙律汁用
- 綜合沙律菜....約 150 克
- 香蕉..............2 隻

  去皮，切成小段
- 雞蛋..............2 個

  焓熟，剝殼，蛋白切粒、蛋黃用叉壓碎
- 中蝦..............10 數隻

  灼熟、剝殼
- 北海道帶子皇 2-3 隻

  先用少許鹽及胡椒粉略醃，大火快煎至兩面金黃半熟，一開四切開
- 火腿..............適量
- 車厘茄..........10 數粒

  一開四切開
- 小黃瓜..........1 條

  切片

沙律汁

- 橄欖油..............約 80 毫升 -100 毫升
- 洋蔥..............1/4 個至半個，視乎大小

  切細粒
- 意大利黑醋..........約 150 毫升
- 留起已壓爛的黑蒜茸 適量
- 鮮榨青檸汁..........1 個份量
- 黑糖舂碎..........適量

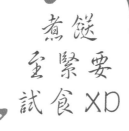

煮餸
至緊要
試食 XD

貼士

1. 此沙律材料亦可隨個人口味，換上其他海鮮、肉類甚至其他時令水果也可以，各俱風味。

2. 黑蒜份量可視乎本身黑蒜的大小，及個人喜愛的份量而作適當調整。

3. 若想賣相美觀一些，蛋黃碎、黑蒜、蝦及帶子可在最後才放上沙律上面，這樣便可以看到更多材料，吸引及美觀一些。

4. 撈好的沙律要盡快進食，否則放久了會出水，大大影響了此沙律原有的新鮮風味。

做法

1. 先用橄欖油將洋葱粒炒香至軟身，盛起，待涼卻後與其他沙律汁材料拌勻成沙律汁，待用。

2. 將全部材料（除火腿及蛋黃外）放入沙律盆，倒入沙律汁撈勻，放上火腿然後盛起上碟，最後灑上蛋黃碎即可享用。

鮮海膽汁伴北海道帶子皇刺身鮮拆蟹粉

Hokkaido giant scallop sashimi with sea urchin cream sauce, crab roe and crabmeat ...180

有時在家請客，若想做些讓賓客有驚喜，而製作又毋須花很多時間的菜式，這道配搭大膽特別的精巧前菜，相信大家一定會喜歡。用海膽打成的鮮海膽汁，配以北海道刺身帶子及鮮拆蟹粉，不單各樣食材鮮味層次豐富，而且在味道及賣相上，都能帶給每位客人滿足和喜悦。

材料

- 北海道特大帶子........12 隻
- 海膽...................約 1 板
  最後放上帶子面
- 三文魚子.............適量

海膽汁

- 海膽...................180 克
  約 4 板
- 鮮奶...................180 毫升
- 鮮忌廉................50 克
- 蠔油...................1 湯匙

蟹黃醬

- 鮮拆蟹粉連蟹肉....180 克
- 薑米...................適量
- 雞湯...................少許
- 鹽......................少許
- 生粉水................少許
  芡用

調味料

- 海鹽...................適量

做法

1. 帶子解凍後用廚紙吸乾水分，放入雪櫃，備用。

2. 海膽汁材料全部放入攪拌機打至幼滑，再用一個較密的隔篩過濾雜質，用保鮮紙封好放入雪櫃保存，備用。

3. 燒熱少許油，將薑米及鮮拆蟹粉連蟹肉炒香，下雞湯及鹽，再用生粉水埋薄芡成蟹黃醬。

4. 準備一個有深度的盛器，先倒入海膽汁，將帶子一開四，放入盛器，再依次放入適量海膽、蟹黃醬及三文魚子，灑上少許海鹽，最後加上裝飾即可享用。

貼士

1. 所有材料（除已製作好的蟹黃醬）在製作前必需放在雪櫃，除了保鮮及衛生外，冰凍進食，更添風味。如條件許可，餐具也最好先放入雪櫃雪凍，效果更佳。

2. 除了配以蟹黃醬之外，配以黑松露醬、XO 醬等效果也不錯，各有風味。

加啲鹽，
加啲糖，又要試味，
又要改稿，
狠忙喔！

香蒜刁草鮮露
欖油鮮蠔漬

Soy-poached oysters marinated in dill and olive oil...177

蠔的吃法真的很多，生、半熟以至全熟都有它的不同風味，這道漬生蠔，經煮製過後再配以橄欖油及香料醃製，在鮮美之間多了一份平時吃蠔少有的質感及香氣。如宴客時做頭盤，此菜可預先做好，到上菜時連玻璃瓶一起奉上，賣相別致之餘，更可節省你在廚房的時間，可以讓出多些時間招呼客人。

材料

- 生蠔 .... 2 打
  24 隻約 400 克
- 小辣椒乾 .... 數隻或適量
- 香葉 .... 4-5 片
- 蒜頭 .... 3-4 粒
- 新鮮刁草 .... 30 克
- 橄欖油 .... 約 250 毫升
  要可蓋過全部生蠔

調味料

- 日本清酒 .... 3 湯匙
- 味醂 .... 2 湯匙
- 家樂牌鮮露 ...... 2 湯匙
- 砂糖 .... 1 1/2 湯匙

貼士

1. 每次買回來的生蠔大小都不一,因此生蠔與橄欖油的份量比例,要以可浸過煮熟的蠔為準。

2. 製作此菜一定要買新鮮的生蠔,否則煮出來會帶有腥味。

做法

1. 小辣椒乾及香葉抹乾淨,蒜頭切片,刁草略切,調味料拌勻,起出蠔肉,備用。

2. 將生蠔放入易潔鑊中,以中火煮開,此時生蠔會出水,輕輕翻動生蠔,煮至水分收乾,此時生蠔質地會變得略為結實;加入調味料炒勻,繼續煮至蠔汁收乾,熄火,撈起蠔,待涼備用。

3. 取一玻璃容器,放入已涼卻的蠔,均勻放入乾辣椒、香葉、蒜片及刁草,最後倒入橄欖油,放入雪櫃泡醃兩天後便可食用。

海膽豉油凍豆腐

Cold tofu appetizer dressed in sea urchin soy sauce...181

在家享受下廚樂或宴請親朋，若能以簡單方法，又可做出美味又讓人驚喜的菜式，都是每個煮人非常期盼又享受的事。這道「海膽豉油凍豆腐」，做法既簡易，賣相又美觀名貴，而且味道非常鮮美好吃，彷彿就讓你的賓客置身高級日本餐廳一樣。

這款海膽豉油汁，原本是朋友從日本帶回來送給我的手信，我試過後覺得非常特別，而且味道鮮美，於是，我便嘗試自己試做。試製成功後，我覺得這海膽豉油還比朋友從日本帶回來的更鮮美，原因是使用新鮮材料，即做即吃，不是長時間保存及沒有防腐添加物。

而且我還發現，這海膽豉油非常好用，除了配凍豆腐外，撈麵、撈飯，作為火鍋的醬汁，蘸湯渣，淋在烤海鮮等等都非常美味！下次下廚，大家不妨感受一下。

材料

- 日式絹豆腐、木棉豆腐或自己喜愛的凍豆腐.................1 件
- 海膽.................適量
- 鮮拆蟹肉.................適量
- 三文魚子.................適量

海膽豉油汁

- 海膽.................80 克
- 魚生豉油.................50 克
- Wasabi.................適量

做法

1. 將海膽、魚生豉油及 Wasabi 放入攪拌機，打至幼滑，倒出，成海膽豉油汁，備用。
2. 取出豆腐放在盛器上，放上適量海膽、鮮拆蟹肉及三文魚子，淋上適量海膽豉油汁及裝飾，即可享用。

貼士

1. 若做好的海膽豉油汁不是即時使用，或是有剩的話，都必須放在雪櫃保存。
2. Wasabi 的份量要視乎個人接受辣的程度而定。
3. 在做此海膽豉油汁時，魚生豉油一定要在做之前放入雪櫃雪凍，這樣才能保持海膽的鮮美風味。

北海道帶子皇沙律

Thai-style
Hokkaido giant scallop
salad...182

泰國菜之中，涼拌沙律菜式佔了一個很重要的部份，要數最受歡迎及最經典的，一定是鮮蝦柚子沙律了，酸甜鮮美，非常開胃。而每間泰國餐廳，她們都會有很多不同版本、不同配搭的涼拌沙律，有的是用新鮮水果配海鮮，有的是混合不同香料及香草配搭不同肉類。但這些美味沙律的主要調味汁都離不開辣椒、青檸汁、魚露以及椰糖，但它們的比例，都會因應每個廚師他們喜歡的口味，或不同材料配搭而有不同的份量。

這道泰式沙律，我選用了北海道的特大帶子皇，因為它鮮嫩味美的質感與泰式的酸甜辛辣非常合襯，而且簡單易做，味道非常清新惹味！

材料

- 北海道鮮帶子.........5 隻
- 香茅.........2 枝
  只要白色部份
- 蒜頭.........2-3 粒
- 芫茜.........2 棵
- 泰國番茄.........約 6 粒
- 柚子肉.........適量
- 指天椒.........1-2 隻
  視乎個人愛辣的程度
- 金不換.........1 棵
- 乾葱頭.........3 粒

醃料

- 鹽及胡椒粉.........各適量

沙律汁

- 青檸汁.........3 湯匙
- 魚露.........2 1/2 湯匙
- 泰國椰糖.........2 湯匙
- 砂糖.........1 湯匙

做法

1. 帶子解凍後用廚紙吸乾水分,再用適量鹽及胡椒粉略醃;香茅切薄片,蒜頭切片炸脆,芫茜切成小段,泰國番茄一開四,指天椒切圈,金不換只要葉,乾葱頭切薄片;沙律汁材料攪勻至砂糖完全溶解,調成沙律汁,備用。

2. 帶子先煎至兩面金黃,微焦帶脆,一開為二,除炸蒜片外,將所有材料放入沙律盆中,倒入沙律汁,快速撈勻,上碟,最後在沙律上面灑上炸蒜片即可享用。

貼士

1. 煎帶子時一定要用大火,快速地將兩面煎至焦香,這樣帶子才可以做到中間還是生的狀態,吃起來才會外脆內嫩,以及可以品嘗到北海道帶子的鮮嫩美味。

2. 此沙律除了帶子之外,還可以其他自己喜歡的海鮮代替,如鮮蝦、魷魚,甚至其他肉類,各有風味。

3. 炸蒜片要最後才灑上,因為如果蒜片連同其他材料及沙律汁一起撈的話,蒜片便會變濕,大大影響了炸蒜片的香脆度。

廚房的零失敗幫手
焗爐

我有好多零失敗的菜式,都是用焗爐炮製,如燒雞、薄餅、蛋糕等等。所以有些初入廚又想自己煮番幾味的朋友來問我的意見,哪種煮食用具最能令他們零失敗,用途又多樣化的,我多會推薦焗爐。

調校好溫度、時間,將已處理、調味的食物放入焗爐,期間你可以烹調其他菜式、上網打機等等,在預定的時間就有香噴噴的東西吃,多方便呢!

話梅花雕醉大閘蟹

Drunken hairy crabs with Shaoxing wine and dried plums...183

私房菜就是沒有固定餐牌，廚師只會按當時的時令食材烹調菜式給客人。所以逛街市已經成為我的習慣，就算到外地旅行，我必定會抽時間去當地市場逛逛，好好去了解當地飲食文化，每次都叫我獲益良多。

在大閘蟹當造的季節，如果你已經吃膩了清蒸的做法，不妨試試這個「話梅花雕醉大閘蟹」！我除了用花雕酒外，更加上口感清香綿長的高粱酒泡醃一晚，香、醇、鮮！保證大家一定囍歡！

材料

- 大閘蟹 .... 約 20 隻
  每隻約六兩
- 紫蘇葉 .... 約十數片

醃料

- 高粱酒 .... 1 公升
- 花雕酒 .... 2.4 公升
- 話梅 .... 150 克
- 砂糖 .... 900 克
- 生抽 .... 1.9 公升
- 老抽 .... 2 湯匙
- 八角 .... 10 粒
- 桂皮 .... 20 克
- 月桂葉 / 香葉 .... 10 片
- 花椒 .... 10 克
- 豆蔻 .... 10 克
- 指天椒乾 .... 10 克
- 葱 .... 60 克
- 蒜頭 .... 8 粒
  拍鬆
- 檸檬 .... 1 個
  切片

> 等大閘蟹涼後才拆繩，就不會燙手。

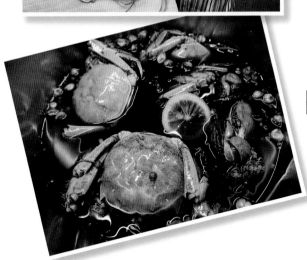

做法

1. 大閘蟹沖洗乾淨，蒸鑊放入紫蘇葉，水滾後放入大閘蟹，以大火蒸 15 分鐘至熟，盛起，待涼備用。

2. 將醃料之香料洗淨，瀝乾水分，將全部醃料放入鍋，煮滾即熄火，待涼備用。

3. 將涼卻的大閘蟹解去草繩，放入涼卻的醃料內，放入雪櫃泡醃一天後，便可盛起享用了。

貼士

1. 經醃料泡醃一天（24 小時）後，大閘蟹已經非常入味，若不是即時食用的話，亦要將大閘蟹撈起，以免泡醃時間過長令味道太濃，蓋過了大閘蟹的鮮味。

2. 煮醃料時，一滾起便要熄火，因高粱酒及花雕酒都不宜長時間加熱，否則酒精濃度便會被揮發，大大影響了原有的風味。而且醃料在冷卻過程中的溫度及時間，已足夠讓香料出味。

3. 泡醃好的大閘蟹放在雪櫃可保存二至三天，但最好還是盡快食用以免變質。

4. 在沒有大閘蟹的季節，亦可以此方法泡醃花蟹或其他蟹，各俱風味。

仁稔醬蒸魚頭

Steamed fish head in Renren sauce...181

讓我們嘗嘗久違了的仁稔,一起尋回小時候的美味吧!

仁稔醬味道微酸,能夠中和食物的油膩感,配以魚頭同蒸,效果極佳,別有一番風味。

材料

- 大魚頭............1個
  約1斤
- 仁稔醬............250克
- 薑片............4-5片
- 芫茜或葱絲....適量
- 油................約3湯匙

調味料

- 砂糖............1/2茶匙
- 生抽............1茶匙
- 花雕酒............1茶匙
- 雞粉............1/2茶匙
- 生粉............1/2茶匙

做法

1. 魚頭洗淨,瀝乾水分;將調味料撈勻,均勻抹在魚頭上。在蒸碟底放上薑片,再放上魚頭,醃約15分鐘,再將仁稔醬鋪在魚頭上。

2. 蒸鑊煲滾水,放入魚頭,加蓋,以大火蒸約10至12分鐘(視乎魚頭之大小),取出,放上適量芫茜或葱絲,將油燒滾淋在魚頭上即可享用。

貼士

- 蒸魚頭時更可在碟底放入適量豆腐卜或陳村粉等吸味食材同蒸,更添風味。

私房陳皮欖豉醬蒸生蠔

Steamed oysters with dried tangerine peel and olive black bean sauce...184

蠔除了刺身生食之外，還可熟吃！甚至還可以品嘗它處於半生熟的獨特鮮嫩質感。這道蒸蠔菜式，以大火短時間將蠔蒸至半熟，讓生蠔表層剛熟，內層還保持生蠔的鮮美，再配以自家製的私房欖豉醬，惹味、鮮味一起共享。

材料

- 美國生蠔 .... 12 隻
- 蔥花 .... 適量

欖豉醬

- 陳皮 .... 1 角
- 豆豉 .... 50 克
- 欖角 .... 50 克
- 蒜頭 .... 約 10 粒
- 指天椒 .... 約 2-3 隻
  視乎吃辣程度
- 油 .... 150 毫升

做法

1. 欖豉醬做法：陳皮先用水浸軟，洗淨，用小刀刮去皮層下的瓤，再切成幼絲。豆豉及欖角洗淨，吸乾水分，切成細粒；蒜頭切成茸，指天椒切圈，備用。鑊內燒熱油，先將蒜茸炒至金黃，再將其他欖豉醬材料下鑊炒香，即可盛起成為欖豉醬。

2. 先用蒸鑊將水煲滾，備用。再將處理好的生蠔放入微波爐以高溫「叮」30 秒，取出，放上適量欖豉醬，馬上放入蒸鑊，加蓋，以大火蒸一分鐘至一分半鐘（視乎蠔之大小及自己喜愛吃蠔的生熟程度），取出，灑上蔥花即可享用。

貼士

1. 生蠔在煮用前一定要放在雪櫃，以確保新鮮及衛生。同時在烹調前先用冰水略為浸洗，確保不會殘留蠔殼，否則吃時咬到蠔殼就大煞風景。

2. 蒸蠔前先用微波爐將蠔「叮」30 秒，目的除了是可鎖住蠔本身的水分之外，還可讓處於低溫的蠔馬上回復室溫，蒸時便可達到表層熟，內層還可保持鮮嫩的質感。若直接將蠔放入蒸鑊蒸的話，蒸的時間相對亦會較長，因而亦會令蠔出水而流失水分，蒸好後的蠔就沒那麼飽滿鮮美。

3. 蒸蠔時一定要先將水煲滾及用大火，蠔一放入便加蓋及計時，若水未滾便放入蠔的話，蠔在鑊內的時間便會太長，蒸出來的蠔便會太熟，失去半生熟的鮮嫩口感。

椰青蛋白 蒸花蟹

Steamed swimmer crabs on egg white and coconut custard...185

海鮮種類繁多，當造期各有不同，在香港這個「美食天堂」，除了可享用本地水域內捕獲的海產外，更有來自世界各地空運到港的海鮮，阿拉斯加的長腳蟹、澳洲的龍蝦、日本的松葉蟹……真是一年四季都能嘗到各地的鮮味。

今次介紹的菜式，用的是本地較常吃到的花蟹，配上清甜的椰青，晶瑩嫩滑的蛋白吸收着蟹與椰青的精華，而且是零失敗的製作，大家務必一試！

材料

- 花蟹 .... 2 隻
  共約 2 1/2 斤
- 椰青水 .... 400 克
- 椰青肉 .... 適量
- 蛋白 .... 280 克
  約 8 個雞蛋的蛋白
- 葱花 .... 適量

調味料

- 鹽及砂糖 .... 各 5 克

做法

1. 蟹劏好、洗淨、斬件，瀝乾水分。蒸鑊煲滾水，以大火蒸蟹約 6 分鐘，取出，將蒸出的蟹水倒出，略放涼一會，備用。

2. 將椰青水、蛋白、蟹水及調味料打勻，倒入一隻有深度的蒸碟，放上蟹件，蓋上保鮮紙，水滾後以大火蒸約 8 分鐘，取出放入椰青肉，再蓋上保鮮紙，以大火再蒸約 2 分鐘至蛋白凝固及蟹熟透，取出，灑上蔥花後即可享用。

貼士

1. 因蒸蟹比蒸蛋白的時間長，因此一定要將蟹先蒸一段時間，才與蛋白同蒸，這樣蟹才夠時間完全蒸熟，而蛋白又可保持着嫩滑的狀態。

2. 新鮮椰青可以在各區一些專賣椰子、香料的店舖買到，但椰青很易變壞，尤其在炎熱的夏季，所以買回來的椰青一定要放在雪櫃貯存。若買不到新鮮椰青，亦可以超級市場的冷凍椰青水飲品代替，效果也不錯，但當然沒有了嫩滑的椰青肉。

3. 椰青肉鮮嫩清甜，不宜太熟食用，所以在蛋白蒸好前兩分鐘才放入略蒸即可，否則蒸得太老了便不好吃。

千層蛋白伴海膽

Egg white omelette mille feuille topped with sea urchin and salmon roe...186

材料

- 蛋白　　　　　300 克
- 　　　　　　　約 8 個
- 鮮奶　　　　　250 克
- 海膽　　　　　約 30 克
- 三文魚子　　　適量

調味料

- 生粉　　　　　25 克
- 水　　　　　　約 2 湯匙
- 鹽　　　　　　4 克
- 砂糖　　　　　3 克
- 雞粉　　　　　2 克

雞蛋的烹調方法繁多，外國甚至有專門介紹雞蛋料理的食譜書。本地家庭最常見的菜式有焓蛋、炒蛋、蒸水蛋、燉蛋等等。這個「千層蛋白伴海膽」，外形似台灣的綿綿冰，口感嫩滑，配上新鮮海膽，令味道更添層次。

這個炒蛋白的方法看似複雜，但只要跟足步驟，大家亦一樣可以做到。當你在家請客時，只要運用這少少技巧，保證能給你的客人大大驚囍！

做法

1. 先用水開稀生粉，再與其他調味料撈勻，備用。

2. 將蛋白及鮮奶放入容器，再加入混合好的調味料，用打蛋器打勻，備用。

3. 取一平底鑊，下適量油以中小火燒熱，然後均勻將油搖勻，令鑊面平均沾有油分，再將多餘的油倒出。

4. 將蛋白漿分次倒入平底鑊，搖勻，令蛋白漿平均分佈在鑊面，形成一層半凝固薄薄的蛋白皮，一直繼續以中小火維持着，直至看見蛋白在開始凝固前，用鑊剷將蛋白從右至左剷起，而形成一摺摺像百摺裙一般的效果，將蛋白放入碟中。重複以上步驟，將煎煮好的蛋白，一層層地重疊在碟上，直至將全部蛋白漿完成為止，隨後在蛋白上均勻放上海膽、三文魚子及裝飾，便可以上枱享用了。

貼士

1. 在蛋白漿加入生粉，可以令蛋白漿「鎖」緊些，令完成後的蛋白不易出水，效果更佳。另煎煮蛋白漿前下適量油，除了令蛋白漿不易黐鑊之外，也可令蛋白做出來的質感光亮些及口感更滑一些。另外，在做法 1 先將生粉用水開稀，與調味料撈勻後才與蛋漿打勻，這樣可使調味料與蛋白漿融合得更均勻些。

2. 因為生粉會沉澱，所以在每次蛋白漿落鑊之前，都要將蛋漿再次打勻，以免蛋白漿出現稀濃不均勻的情況。

3. 要令蛋白做出一層層、一摺摺的理想效果，關鍵在於蛋白一定要在凝固前，在它半凝固的狀態時，將蛋白剷起。否則如蛋白太熟及凝固變硬後，便變成「一塊」煎蛋白，做不成預期的一層層效果了。

4. 在烹煮的全程中，必需要以中小火煎煮，這樣可以更易掌握及控制蛋白的凝固效果，太大火的話，蛋白便很快熟，質地亦會太實，還一定要在蛋白在凝固前，半凝固的時候剷起，這樣做出來的蛋白才有非常嫩滑的效果。

5. 此道菜除了配以海膽、三文魚子外，我還試過配以鮮拆蟹肉、黑松露及各類菇菌，各俱風味，效果也相當不錯。

黑蒜陳皮火腿南棗蒸石斑

Steamed grouper with black garlic, ham and dates...200

小時候父母常說「吃魚有益」，而且魚的煮法簡單，用蒸已能帶出鮮味，所以它確實是我家飯桌的「常規成員」呢！

以往蒸魚，不外乎用薑葱、陳皮、豆豉、麵豉等，視乎魚的種類而定。今次特別用上自家製黑蒜，炮製一道簡單養生的家庭小菜「黑蒜陳皮火腿南棗蒸石斑」。有時配料上的小改變，能為味覺帶來極大的新鮮感呢！

材料

- 石斑............1 條
  約 1 斤至 1 斤半
- 黑蒜............10 粒
  略切碎
- 陳皮............1 小塊
  用水浸軟，洗淨後切絲
- 火腿............約 50 克
  切粒
- 薑............約 40 克
  切粒
- 南棗............約 4 粒
  洗淨、去核、切粒
- 葱............約 3 棵
  切絲

調味料

- 蒸魚豉油......適量
- 油............適量

做法

石斑劏好、洗淨，吸乾水分，放上蒸碟，除葱絲外，將其餘材料平均放在魚身上面，水滾後放入蒸鍋，以大火蒸約 10 至 12 分鐘，取出，放上葱絲，淋上少許蒸魚豉油及潷適量滾油即可享用。

貼士

我試過用龍躉腩來代替石斑，除了是因為啖啖肉之外，龍躉腩那層厚厚充滿骨膠原的魚皮，香綿軟滑，蒸出來的魚油鮮香味與黑蒜的獨特香氣相配，效果亦相當好。

胡椒海鹽焗蟹
配泰惹味薄荷汁

Pepper salt-baked crab
with Thai mint sauce...187

焗爐是我的好幫手，如今次介紹的「胡椒海鹽焗蟹」，只要調校好溫度、時間，將處理好的醃料及蟹一同放入焗爐就可以了，期間可烹調其他菜式，或者招呼朋友。

材料

- 肉蟹.............2 隻
  約共 2 1/2 斤
- 白胡椒.........60 克
- 粗鹽.............550 克

薄荷汁

- 新鮮薄荷葉.....40 克
- 鮮榨青檸汁.....90 克
- 魚露.............45 克
- 砂糖.............80 克

做法

1. 白胡椒舂碎，與粗鹽混合，放入鑊中，以中慢火炒至粗鹽焦黃，散出胡椒香氣，盛起，備用。

2. 蟹劏好洗淨（不用斬件），起出蟹蓋，焗盤鋪上錫紙，放上蟹，蓋上約兩張紗紙，將蟹包裹着，然後將炒好的胡椒鹽均勻地覆蓋在上面。

3. 預熱焗爐攝氏 250 度，將蟹放入焗約 25 分鐘。焗蟹期間，將薄荷汁材料全部放入攪拌機打碎成薄荷汁。

4. 蟹焗好後小心倒出胡椒鹽，拆開紗紙，將蟹取出，斬件，上碟，吃時蘸上適量薄荷汁伴吃即可。

貼士

1. 焗蟹時不要將蟹斬件，其一是若一早將蟹斬件，焗好後會較難將蟹起出上碟；其二，將蟹起出時會有機會沾到焗盤上的鹽，若蟹斬開了，便會有機會沾到太多，蟹肉便會很鹹。

2. 此做法除了焗蟹外，還可以用來焗大蝦、花蛤、魚等，各俱風味，但焗的時間便要作出適當調整。

手撕勁滑葱油雞

Hand-shredded chicken in spring onion ginger dressing...188

中國人宴客菜式當中，「雞」差不多是必然的選擇，因為它的可塑性高，不論焗、燒、炸、燜、蒸……都各俱風味，而且中國人喜歡「有骨落地」，寓意菜餚豐富。

如果大家希望吃時可以更為方便乾淨，不用「落手落腳」的，可以嘗試這個手撕勁滑葱油雞。以低溫烹調（slow cook）方法將雞浸熟，能保留雞肉的嫩滑，用手撕方式起肉，再配搭私房特製葱油醬，每啖鮮嫩雞肉都散發出誘人的葱油香味，保證大家一定囍歡！

材料

- 雞 .... 1 隻／約 2 斤半
- 薑片 .... 8-10 片
- 葱 .... 約 5-6 棵
- 八角 .... 約 2 粒
- 香茅 .... 3 枝
  只用白色部分
- 斑蘭葉 .... 3-4 片
- 鹽 .... 2 湯匙
- 冰水 .... 1 盆
  要蓋過雞

葱油醬

- 葱約 300 克 .... 約 20 棵
- 乾葱頭 .... 約 10 粒／約 80 克
- 鮮沙薑 .... 約 3 粒／約 40 克
- 油 .... 250 毫升
- 海鹽 .... 10 克
- 沙薑粉 .... 5 克
- 雞粉 .... 少許

做法

1. 煲滾水（水要蓋過雞），下薑片、葱、八角、香茅、斑蘭葉及鹽，攪勻至鹽完全溶解，下雞浸約十數秒，馬上取出，以凍水沖約 3 分鐘。

2. 原煲水再煲滾，下雞，至水再滾起，加蓋關火，浸約 45 分鐘，取出雞，放於冰水中浸泡約 20-30 分鐘，取出，瀝乾水分後手撕起肉，上碟。

3. 將葱、乾葱頭、鮮沙薑全部切碎。燒熱油，先下乾葱頭及沙薑碎爆香至散發出香氣，下葱花略炒，下海鹽、沙薑粉及雞粉炒勻，淋上雞肉上即可享用。

貼士

1. 將浸熟的雞馬上泡浸冰水，目的是讓雞急速降溫，停止繼續烹調，從而令雞皮馬上收縮，達至吃時有皮爽肉滑的口感。

2. 雞隻大小及重量不一，泡浸的時間相對要延長或減短一些。亦可以用長竹簽插入雞腿位置，若沒有血水滲出，便代表雞隻經已熟了。

3. 泡完雞的湯，更可以用來煮飯、灼菜、做煲湯的湯底等，好味之餘又能物盡其用。

黑松露
魚茸豆腐

Fried dace tofu strips
in black truffle sauce...189

這是港式經典家常菜「老少平安」
的變奏版，但在處理上，將蒸好的
魚茸豆腐雪硬後切條再煎香，還配
以名貴的自製黑松露醬，無論在賣
相及味道上，都得以進一步的昇華！

材料

- 布包豆腐 .... 3 件
- 免治鯪魚肉 .... 150 克
- 蝦乾 .... 30 克
- 雞蛋 .... 1/2 個
- 生粉 .... 20 克

調味料

- 鹽、砂糖、雞粉 .... 各 1/2 茶匙
- 胡椒粉 .... 少許
- 麻油 .... 少許

黑松露醬

- 冰鮮黑松露 .... 35 克
- 橄欖油 .... 適量
- 鹽 .... 約 1/4 茶匙

生粉漿：煎魚茸豆腐時用

- 生粉 .... 25 克
- 雞粉 .... 3 克
- 水 .... 約 30 克

做法

1. 布包豆腐用篩隔着，用手搓爛，滴出的水不要（因太濕的話，會令做出來的效果沒有黏性，會散）；蝦乾用水浸軟，瀝乾水分，用鑊炒至捲起及散出香氣，瀝乾油分，剁碎，備用。

2. 搓爛的豆腐與鯪魚肉、蝦乾碎、雞蛋、生粉及調味料撈勻，攪至起膠及有黏性。

3. 取一方形淺深度的蒸盆，搽上一層薄薄的油，將攪好的豆腐放入，鋪平，蓋上保鮮紙，水滾後以大火蒸約 18 至 20 分鐘（視乎豆腐的厚薄）成魚茸豆腐。

4. 將黑松露剁碎，加入橄欖油及鹽撈勻成黑松露醬。將生粉漿材料撈勻，開成糊狀，備用。

5. 蒸好的魚茸豆腐待涼卻後，放入雪櫃冷藏約半天令其定形。

6. 從雪櫃取出魚茸豆腐，切成條狀，沾上適量生粉漿，以中火煎至兩面金黃香脆，上碟，放上適量黑松露醬，加上裝飾，即可享用。

貼士

1. 在魚茸豆腐內加入炒香的蝦乾碎，目的是可以增加魚茸豆腐的香氣，蝦乾除了以炒處理外，還可以用焗爐烘香或炸都可以。

2. 煎魚茸豆腐前，在魚茸豆腐四周表面沾上生粉漿，可令煎出來的魚茸豆腐表面更加香脆。

3. 若做傳統蒸的「老少平安」，可以選用現成包裝蒸煮用的豆腐，效果會滑些。但做此菜，我建議要選用在街市買的布包豆腐，因為密度高些，做出來的魚茸豆腐會硬身些，不易散。

海鮮芝士焗牛油果

Avocado au gratin with seafood and cheese...190

材料

- 熟牛油果 .... 3 個
- 中蝦 .... 6 隻
- 鮮帶子 .... 5-6 隻
- 鮮魷魚 .... 約 100 克
- 草莓 .... 6-8 粒
- 青蘋果 .... 1/2 個
- Parmesan 芝士粉 .約 25 克
- Mozzarella 芝士 .約 120 克
- 沙律醬 .... 約 100 克
- 三文魚子 .... 適量

海鮮醃料

- 鹽及胡椒粉 .... 各少許

在炎炎夏日，一些以水果、海鮮為主要材料的菜式特別受歡迎，既清新開胃，感覺又健康。這道海鮮芝士焗牛油果，做法既簡易，賣相又悅目，美味就更加不用置疑啦！

做法

1. 牛油果洗淨，抹乾水分，一開二，起出中間的核；海鮮、草莓及蘋果全部洗淨，吸乾水分，全部切粒，備用。

2. 海鮮粒加入少許鹽及胡椒粉撈勻略醃，鑊中燒熱油，將海鮮炒至約 6 至 7 成熟，隔走水分，備用。

3. 待海鮮稍為涼卻後，加入草莓粒、蘋果粒、Parmesan 芝士粉、少許 Mozzarella 芝士及沙律醬撈勻成海鮮沙律，然後平均釀入牛油果至適當高度，最後將餘下的 Mozzarella 芝士鋪在海鮮沙律上面，放上焗盤，備用。

4. 預熱焗爐攝氏 200 度，放入牛油果，焗約 10 至 12 分鐘，至 Mozzarella 芝士溶化及表面微焦金黃，便可取出，在上面放上三文魚子及裝飾後，即可上碟享用了。

貼士

1. 海鮮粒不要炒得太熟，否則焗時會令海鮮變得太熟，令肉質變實不好吃。另外炒完海鮮後，海鮮粒排出的水分要隔掉，否則會令到完成後的沙律太濕，影響效果。

2. 牛油果切開後，可在底部約 2mm 位置平切一刀，令牛油果可以平穩地放在焗盤及碟上。

3. 在沙律醬內加入 Parmesan 芝士粉，可以令焗出來的海鮮沙律散發出更香的芝士味，再加入少許 Mozzarella 芝士，可以令焗出來的海鮮沙律更帶有芝士的黏性。

4. 亦可隨個人喜愛，換以其他的海鮮或生果，各俱風味。

冬蔭功脆皮燒雞

Tom Yum roast chicken...191

材料

- 雞 .... 1 隻
  約 2 斤

醃料

- 冬蔭功醬 (a) . 90 克
- 冬蔭功醬 (b) . 40 克
- 蒜茸 .... 25 克
- 乾葱茸 .... 25 克
- 南薑 .... 20 克
  切碎
- 芫茜 .... 2 棵
  整棵連頭，切碎
- 香茅 .... 2 枝
  只要白色莖部，切碎
- 檸檬葉 .... 8-10 片
  切碎
- 水 .... 50 毫升
- 油 .... 3 湯匙
- 生粉水 .... 適量
  埋芡用

調味料

- 鹽 .... 1/4 茶匙
- 砂糖 .... 1/2 茶匙
- 雞粉 .... 1/2 茶匙

冬蔭（即泰式酸辣湯）一般都是配海鮮，例如泰國的經典名湯「冬蔭功」（酸辣蝦湯），後來試過冬蔭雞湯，發覺這個配搭也很特別，於是便創作出這道冬蔭雞的變奏版「冬蔭功脆皮燒雞」。 冬蔭的獨特香辣與脆嫩 juicy 的燒雞是個夢幻組合！再以手撕方法上碟，讓你能夠單單用一對筷子就能體會以往只有在喝冬蔭功湯才能品嘗到的獨有酸、辣、甜、香味道。

做法

1. 雞洗淨,用廚紙吸乾水分,先將冬蔭功醬 (a)90 克均勻抹在雞身內外,備用。

2. 鍋中燒熱油,先爆香蒜茸及乾葱茸,至開始焦黃及散發出香氣,再下南薑、芫茜、香茅及檸檬葉炒香,下冬蔭功醬 (b)40 克炒勻,加水,下調味料炒勻,再用生粉水埋芡,盛起,待完全涼卻後釀入雞肚內,以廚房用鵝尾針縫好,再用保鮮紙包好放入雪櫃醃一晚。

3. 焗爐預熱攝氏 150 度,將雞放入爐焗 60 分鐘後,將溫度升至攝氏 250 度再焗 15 至 20 分鐘,至雞皮金黃香脆。

4. 將雞取出,手撕起肉,然後將雞肉與肚內的醃料撈勻伴吃即成。

貼士

1. 燒雞要做得皮脆肉嫩,一定要揀選夠肥多脂肪的雞,因為在長達個半小時的烤焗過程中,需要有足夠的皮下脂肪可以溶解,令雞皮更加香脆。太瘦的雞燒出來會太乾及口感很「老」。

2. 醃雞前要盡量將雞內外的水分吸乾,否則水分太多的話,會稀釋醃料的濃度,影響了燒雞的風味。

3. 冬蔭功醬是用來煮泰式冬蔭功湯的醬料,在一般專賣泰國食材的雜貨店均可買到,但因為有很多不同的牌子,它的酸辣鹹程度都各有不同,所以要試多幾款選一款合你自己口味的。

4. 醃料在埋芡時不妨可以埋得「傑」身些,這樣會較容易釀入雞肚內。

金不換辣酒煮翡翠螺

Atlantic green whelks in spicy wine sauce with Thai basil...192

每天的工作都忙得不可開交，早前得到一天珍貴的假期，便帶同「叮叮」（我家的法國老虎狗）外出遊玩。悠悠的走到街上，抬頭一看，啊！原來夏天已經到了！街道兩旁的花綻放得燦爛，樹木亦很茂盛翠綠。

炎炎夏日，卻令人懶懶洋洋。這個香辣惹味的菜式，希望能為大家的晚飯桌上帶來小小的新鮮刺激，吃罷元氣滿滿的，繼續接受新挑戰。

材料

- 急凍翡翠螺 .... 約 3 斤
  1.8 公斤
- 蒜茸 .... 60 克
- 乾葱茸 .... 60 克
- 薑米 .... 100 克
- 新鮮指天椒碎 ...... 20 克
- 雞湯 .... 1 公升
- 乾指天椒粉 .... 20 克
- 香葉（月桂葉）.... 10 數片
- 甘草 .... 約 6 片
- 日本葛絲 .... 適量
- 金不換 .... 5-6 棵
  只要葉

調味料

- 紹興酒 .... 500 毫升
- 玫瑰露酒 .... 200 毫升
- 辣椒油 .... 80 克
- 桂林辣椒醬 .... 60 克
- 麻油 .... 70 克
- 沙爹醬 .... 50 克
- 魚露 .... 80 克
- 砂糖 .... 2 茶匙

做法

1. 翡翠螺解凍後洗淨，鍋中下水（水份量要蓋過翡翠螺），凍水下翡翠螺，水滾後煮約30分鐘，熄火，不要開蓋，至水涼卻後將螺撈起，摳去螺頭上的奄，備用。

2. 鍋中下油約4至5湯匙，燒熱，爆香蒜茸、乾葱茸及薑米，下新鮮指天指碎炒至散發出香氣，下雞湯、乾指天椒粉、香葉及甘草，大火煮滾後改以中小火煮約15分鐘（不用加蓋）使香料出味。

3. 加入所有調味料，煮滾後繼續以中小火煮約5至10分鐘，目的是讓部分酒精揮發，而5至10分鐘烹煮時間的長短，則視乎個人喜愛酒味的濃度而決定。

4. 加入翡翠螺，待湯汁煮滾後便可熄火，待涼卻後放入雪櫃貯存讓它入味，備用。

5. 隔日食用時，預先將葛絲浸軟，用滾水煮至透明軟身，先放在盛器底部。將翡翠螺連湯汁翻熱，最後加入金不換撈勻，煮至金不換軟身後便可倒入盛器，上枱享用了。

貼士

1. 此菜香辣惹味，非常之好佐酒，但亦可因應個人喜愛吃辣的程度，而將辣椒香料的份量增多或減少。

2. 此道菜的煮法除了用翡翠螺之外，還可以蝦、蟹或花蛤等代替，各俱風味。但在做法4將蝦、蟹或花蛤煮熟後，便可以馬上趁熱享用，毋須放過夜讓它入味了。

3. 此湯汁非常惹味，將它變成火鍋湯底也是一個非常好的「煮」意。而剩下的湯汁，除了配葛絲外，還可配以各款即食麵、粉絲、河粉等麵食，非常「索」味好食。

秘製番茄咖喱爛牛肉

Top-secret tomato beef curry...194

咖喱菜式是很多人的至愛，由於它可塑性非常高，無論配搭任何肉類、海鮮甚至蔬菜都非常美味。咖喱種類繁多，煮法亦有很多變化，這道咖喱食譜選用黃咖喱醬，但加入了我很喜歡的番茄口味，濃郁的香、辣之間增添了一份番茄獨有的酸甜風味，無論配飯、麵、薄餅、包類都非常醒胃好吃！保證喜歡咖喱的朋友一定會囍歡！

材料 A

- 牛腱 .... 1.5 千克
- 洋葱 .... 2 個
- 薯仔 .... 約 3-4 個
  去皮、切角
- 罐頭去皮番茄 ...... 500 克
  略為切碎
- 新鮮番茄 .... 3 個
  切件

材料 B

- 黃咖喱醬 .... 450 克
- 牛肉原湯 .... 約 1.5 公升
- 香茅 .... 3 枝
  拍鬆

調味料

- 魚露 .... 80 克
- 砂糖 .... 50 克
- 茄汁 .... 200 克
- 椰汁 .... 約 1/2 杯

做法

1. 牛脹洗淨，飛水，原件加水約 2 公升先煲約 1 小時，撈起，涼卻後切成適當大小，煮牛脹的牛肉原湯留用。

2. 洋葱切粒，薯仔煎香備用。

3. 鍋中下油約 3 湯匙，燒熱，下洋葱炒香至軟身，加入咖喱醬炒香，下牛肉原湯、罐頭去皮番茄、香茅、魚露、砂糖、茄汁及牛脹，以慢火燜煮約半小時後，下薯仔再燜煮 15 分鐘，最後下新鮮番茄，繼續以小火再燜煮約 15 分鐘後熄火，讓它自然涼卻。隔日翻熱食用風味更佳，吃時加入椰汁煮滾或直接倒入椰汁拌吃均可。

貼士

1. 用牛脹燜煮湯汁較濃的菜式時，最好先將牛脹煲約 1 小時，讓它煮至半鬆軟。目的是在調味後正式燜煮時，可減短燜煮的時間，否則湯汁會愈煮愈濃及愈煮愈「傑」，難以控制效果及味道。

2. 烹煮番茄味道的菜式時，除了新鮮番茄之外，加入罐頭番茄及茄汁，可令番茄的味道更突出及香濃。

3. 黃咖喱醬可於一般賣香料椰汁的店舖或超市都可以買到。

罐去皮番茄
去咗邊度呢？

大牌檔滷豬肉蛋豆腐

Marinated pork, eggs and tofu in "Dai Pai Dong" style...195

材料

- 五花腩 .... 2 斤
- 蒜頭 .... 約 8-10 粒
  拍鬆
- 炸豆腐 .... 5 件
- 焓熟雞蛋 .... 8 個
  去殼

滷水料

- 八角 .... 4 粒
- 桂皮 .... 1 小段
- 白胡椒 .... 1 湯匙
- 芫茜頭 .... 8 棵
  拍鬆
- 指天椒或辣椒乾 ... 3-5 隻
  視乎個人愛辣的程度
- 黑豉油 .... 6 湯匙
- 魚露 .... 8 湯匙
- 椰糖 .... 60 克
- 水 .... 8 杯

相信大家都去過泰國旅行吧，在很多夜市的大牌檔小攤子、商場的 Food Court 裏面，大家都一定會見過或吃過這道泰式滷肉。我相信這是最經典的泰式碟頭飯之一，它有點中國菜的風味，但又帶點香辣的泰菜味道，很濃郁惹味，絕對是飯、麵、粥品的最佳配搭。

我最愛的反而是配料中的滷蛋及豆腐，盡吸豬肉及滷汁的精華，啖啖美味。其實這道菜不難做，而且包保一定成功，大家不妨在家試試，吃得開心！

做法

1. 五花腩洗淨，切成約寸半大小的方形，吸乾水分，備用。

2. 鍋中燒熱油約 3 湯匙，先用小火將蒜頭炒至金黃色，再下豬肉，加大火，將豬肉炒至變成白色及微微焦香。備用。

3. 將滷水料的香料放入小湯袋中，綁好，與其他滷水料的材料、五花腩、炸豆腐及焓雞蛋加入鍋內，大火煲滾後改小火煮約 1 小時 30 分鐘（不用加蓋），至五花腩軟腍鬆化。

4. 如馬上食用的話，將火加大，將滷汁略為煮「傑」後便可關火，上碟享用了。另一吃法是，當豬肉煮好關火，整鍋涼卻後擺放一夜，隔天翻熱食用風味更佳。

貼士

1. 黑豉油是東南亞常用的醬料，深黑色，帶有一種焦香的甜味，較一般豉油「傑」身及豉油味較濃，通常用作佐料醬汁用途，如吃海南雞飯，蘸湯渣等，亦可用作炒、煮、燜、滷及涼拌醬汁的調味。在一般賣椰汁香料店舖、泰國食品舖或一些大型超市都可以買到。

2. 煮這道菜全程都不用加蓋，目的是要令滷汁越煮越濃稠，令煮出來的滷肉及滷汁味道更豐厚。

3. 八角及桂皮等香料先放入白鑊，用小火燒至散出香味後燜煮，味道會更香。

泰好味福食燒雞
Thai-style roast chicken...196

我們餐廳的員工飯桌，是廚師們練習廚藝、發揮創意的「武道場」。每當私房菜有任何新菜式，最初一定會在福食中出現，讓同事們試試，俾意見！所以，如果你發現我們的福食中有北海道帶子皇、老虎蝦、龍蝦等菜式，請不要太驚訝！！

今次這道「泰好味福食燒雞」是我們餐廳的一位泰籍華人廚師的家傳秘製燒雞，真的「狠姣痴」（很好吃）！誠意推介大家在家試做！

材料

- 雞 .... 1 隻約 2 1/2 斤

醃料

- 蠔油 .... 2 1/2 湯匙
- 麻油 .... 2 茶匙
- 美極鮮醬油 .... 2 茶匙
- 五香粉 .... 1/2 茶匙
- 泰國椰糖 .... 2 茶匙
- 雞粉 .... 1 茶匙
- 黑胡椒碎 .... 1 1/2 茶匙

  舂碎
- 泰國蒜頭 .... 5-8 粒

  連皮舂爛
- 芫茜頭 .... 3 棵

  舂爛

泰式炒米碎

- 糯米 .... 60 克
- 檸檬葉 .... 3-4 片

  摙去葉中間硬條
- 香茅頭 .... 1 枝
- 南薑 .... 10 克

燒雞汁

- 泰式辣椒碎 .... 約 2 茶匙
- 泰國椰糖 .... 2 茶匙
- 魚露 .... 1 1/2 湯匙
- 青檸汁 .... 1 湯匙
- 酸子汁 .... 2 茶匙
- 炒米碎 .... 1 茶匙
- 芫茜、葱、薄荷葉、乾葱頭

  .... 各少許，切碎

做法

1. 泰式炒米碎：將檸檬葉、香茅頭及南薑切碎，與糯米混合，用白鑊以小火烘至啡色焦燶及散發出香氣，盛起，涼卻後舂碎即成。

2. 將雞洗淨開邊，用廚房紙抹乾水分，將醃料拌勻，均勻搽滿雞身內外，用保鮮紙包好，醃一晚。

3. 將燒雞汁拌勻備用。預熱焗爐約攝氏185度，將雞放入（皮向上）燒約40至45分鐘至雞皮金黃香脆熟透，便可取出斬件或手撕，伴以燒雞汁享用。

貼士

1. 在燒雞前將雞腿向下拉一下，可令雞肉鬆弛一些，燒起來亦較均勻易熟。

2. 此燒雞除以焗爐烤焗製作外，還可以炭火燒烤，或將雞斬件醃好再上生粉炸亦相當好食，如貪方便，甚至可以全雞翼製作，各有風味。

3. 此米碎是泰國菜常用的香料之一，可用於泰式醬汁、泰式生茶包或沙律，倍添風味。

韓風芝士泡菜帶子燒
Kimchi scallop roll au gratin...197

肥美肉厚的北海道特大帶子，是很多喜歡吃日本菜朋友的至愛，除了刺身蘸豉油青芥辣、配魚生飯之外，還可以香煎或燒烤，味道非常鮮嫩味美。這道食譜以低溫慢煮方式處理，口感徘徊在生與熟之間，再配以韓國泡菜及芝士，在鮮嫩的味道上增添了更複雜的層次和質感，為味蕾的觸感帶來一次全新體驗，而且製作簡單，絕對是在家宴客零失敗的美味之選！

材料
- 北海道刺身帶子......6 隻
- 原棵韓國泡菜........1 棵
- 車打芝士.............適量
- 三文魚子.............適量
- 烘香的煙肉碎........適量
- 鹽....................約 2 茶匙

工具
- 廚房用溫度計
- 廚房用火槍

做法

1. 帶子先解凍，鍋中注滿約 2/3 水，放入鹽攪勻，放入溫度計，以小火將水煮至約攝氏 70 度，放入帶子，浸約 2 至 2 1/2 分鐘，馬上撈起浸冰水約 2 分鐘，撈起，用廚紙吸乾水分。

2. 將泡菜鋪平，放上帶子，然後用泡菜將帶子整隻包好，上碟，將適量車打芝士刨或切碎，鋪在泡菜上，再用火槍慢火將芝士燒至半溶，放上三文魚子及灑上適量烘香的煙肉碎即可享用。

貼士

1. 北海道刺身帶子本身已經可以生食，口感鮮嫩，所以在水中泡浸的時間不宜太長，過熟的話，便會失去鮮帶子應有的鮮嫩質感。

2. 在用泡菜包帶子之前，帶子一定要用廚紙吸乾水分，否則若帶子水分太多的話，包出來便會太濕而且水分便會由泡菜滲出，大大影響了賣相及味道。

3. 北海道帶子可於大型日式超市或日本食品店均可以買到。

香脆韭菜煎蠔餅

Chive oyster fritter...198

一般蠔餅都是以較便宜及細小的蠔
仔製作，但有時自己下廚或宴客開
飯，當然應該要吃得好些！這道蠔
餅就以美國桶蠔製作，配以香氣撲
鼻的韭菜，做出來的蠔餅吃起來更
鮮嫩肥美，質感飽滿，而且更是零
失敗製作，吃得開心！

材料

- 新鮮美國桶蠔..........700 克
- 韭菜....................300 克
- 番薯粉..................150 克
- 麵粉....................150 克
- 雞蛋....................2 個
- 水......................約 600 毫升

調味料

- 鹽......................1 1/2 茶匙
- 胡椒粉..................適量

做法

1. 生蠔洗淨，瀝乾水分，切成適當大小。韭菜洗淨，瀝乾水分，切粒，備用。

2. 先用水撈勻番薯粉、麵粉、雞蛋、韭菜及調味料，最後加入生蠔粒輕輕攪勻成蠔餅粉漿。

3. 用一平底易潔鑊，燒熱油，倒入適量蠔餅粉漿，將蠔餅煎至兩面金黃香脆，盛起，放在吸油紙上吸去多餘油分後，便可上碟享用了。

4. 吃時可隨個人喜愛配上辣椒醬、魚露或 XO 醬，更添風味。

貼士

1. 桶裝生蠔一般較為新鮮乾淨，只需用水略為沖洗便可，但若是買街市散裝蠔的話，便一定要清洗乾淨，大家可以先將生蠔放在盆內，加入適量生粉用手輕輕搓揉，生蠔的污垢便會被生粉吸出黏付着，再用水沖洗乾淨便可使用了。

2. 生蠔不要切得太細粒，煎出來的蠔餅便可以咬到生蠔的鮮嫩質感了。

3. 煎此蠔餅不妨可以下多一些油，以半煎炸的形式處理，這樣煎出來的蠔餅才夠鬆化香脆。另外不要用小火，否則粉漿便會吸收油分，食時會感覺很油膩了。

4. 粉漿調好後才加入生蠔，一來保持生蠔粒的原整性，二來以免生蠔攪拌過度而變得太爛。

5. 水的份量可以因應個人喜歡蠔餅的稀濃程度而加多或減少些。

辣妹水煮牛肉

*Poached beef
in Sichuan "Shuizhu" style...199*

「水煮」是四川菜的一種經典烹調方法，不過傳統的做法頗為複雜、工序繁多，所以不容易在家做。

這次分享的菜式，會用坊間買到的麻辣醬，配上新鮮牛柳、非常好索味的日本葛絲、爽口的小黃瓜……麻、辣、鮮、爽！更是零失敗的簡易菜式，你一定要試試。

材料一

- 蒜頭............4 粒
  拍鬆
- 薑............40 克
  切片
- 三五麻辣料............150 克
- 四川豆瓣醬............80 克
- 辣椒乾............20 克
- 花椒粒............12 克
- 雞湯............1 公升
- 牛油............70 克

材料二

- 日本葛絲............50 克
- 新鮮牛柳............500 克
- 生粉............20 克
- 水............30 克
- 豬紅............300 克
- 大豆芽............100 克
- 小黃瓜............適量
  切片

調味料

- 紹興酒............10 克
- 雞粉............1 茶匙
- 砂糖............3 茶匙

配料

- 炒香芝麻............適量
- 芫茜............2-3 棵
  切段

做法

1. 葛絲用清水泡浸一晚，備用。

2. 牛柳切片，用生粉開水拌勻，與牛柳撈勻備用。

3. 燒熱鑊下少許油，把蒜頭、薑片爆香，放入10克辣椒乾、四川豆瓣醬、三五麻辣料及6克花椒粒，爆香後灒紹酒，最後放入雞湯、雞粉和砂糖煮滾。

4. 煮滾後放入豬紅煮3至4分鐘，再放入大豆芽、小黃瓜及葛絲，煮滾約2至3分鐘後，小心將材料隔起，放上盛器，留起湯汁，備用。

5. 將湯底再滾起，熄火，放入已用生粉水拌勻的牛柳浸煮至約七成熟，撈起，連同湯汁倒入盛器內。

6. 最後將餘下的10克辣椒乾和6克的花椒粒用油爆香，趁熱淋在盛器內的牛柳上，灑上炒香芝麻及放上芫茜裝飾即可享用。

貼士

1. 三五麻辣料是一種現成的濃縮麻辣湯包，很適合用於製作四川水煮及火鍋菜式，非常方便好用，在街市一般南貨店均可以買到。

2. 此菜做法除了用牛肉製作外，還可以用魚、田雞、其他肉類或海鮮等代替，各俱風味。

3. 此道菜的麻辣程度亦可以因應各人吃辣的程度而作適當調整。

泡菜年糕脆皮燒雞

Roast chicken stuffed with Kimchi and Korean rice cake...200

近年韓風在香港盛行，學韓語、追韓劇、聽 K-POP 等等……由娛樂旅遊、電子產品到飲食文化都備受港人吹捧！韓國料理以清淡為主，韓國人認為飲食不只為充飢，更相信食物有其療效作用，固經常使用一些對身體有益的蔬菜、藥材等作食材。所以韓國菜在著重健康養生的香港流行起來，真的不無道理。

「泡菜年糕脆皮燒雞」極具韓國風味，年糕是韓國人的主食之一，而泡菜更是飯桌上不可少的配菜。將年糕、泡菜釀入塗滿韓式辣椒醬的雞中，再燒至香脆，簡單幾個步驟，就能在家炮製創意又有韓國特色的菜式了！

材料

- 雞 .... 1 隻 約 2 1/2 斤
- 韓國辣椒醬 ..... 約 70 克
- 韓國年糕 .... 80 克
- 韓國泡菜 .... 220 克
- 鹽 .... 約 8 克

做法

1. 雞洗淨，吸乾水分；韓國年糕用水浸軟，備用。

2. 將韓國辣椒醬均勻抹滿雞身內外；泡菜切碎，與鹽、年糕撈勻，塞入雞肚，用廚房用鵝尾針縫好，再用保鮮紙包好放入雪櫃醃一晚。

3. 焗爐預熱攝氏 150 度，將雞入爐焗 60 分鐘後，將溫度升至攝氏 250 度再焗 15 至 20 分鐘，至雞皮金黃香脆，即可享用。

貼士

1. 燒雞要做得皮脆肉嫩，一定要揀選夠肥多脂肪的雞，因為在長達個半小時的烤焗過程中，有足夠的皮下脂肪可以溶解，令雞皮更加香脆。太瘦的雞燒出來會太乾及口感很「老」。

2. 醃雞前要盡量將雞內外的水分吸乾，否則水分太多的話，會稀釋醃料的濃度，影響了燒雞的風味。

豉椒蒸排骨
~ 專業廚師家常菜的小竅門
Steamed pork ribs in chilli black bean sauce...201

材料

- 新鮮腩排........1 斤

豉椒醬

*此豉椒醬如蒸一斤排骨
份量的話，約可用兩次

- 豆豉.................80 克
- 指天椒.............1-2 隻
- 蒜頭.................3 粒
- 乾葱.................2 粒
- 陳皮.................1 小角
- 砂糖.................1 茶匙

調味料

- 生粉.................1 茶匙
- 鹽.....................1/3 茶匙
- 砂糖.................3/4 茶匙
- 雞粉.................1/2 茶匙
- 蒜油.................約 2 茶匙

豉椒蒸排骨是一道非常經典的廣東家常小菜，但一般在家製作都是將豆豉等調味料直接與排骨撈勻醃製，製作比較簡單，所以你會覺得在家做出來的蒸排骨，總會不及在出面酒家、餐廳做出來的好吃。今次跟大家分享一下，酒家專業廚師製作此道家常小菜的專業方法，當然，步驟是多了一些，但懂得了這些小竅門，在家做出來的蒸排骨，都可以有專業水準，好吃得多了。

做法

1. 腩排洗淨吸乾水分。豆豉洗淨，吸乾水分。指天椒切圈；蒜頭、乾葱頭剁成茸。陳皮用水浸軟，洗淨，刮去瓤，切絲，備用。

2. 鑊中下約 3 湯匙油，燒熱，先爆香指天椒、蒜茸、乾葱茸及陳皮絲，炒至散發出香氣及微微焦黃，下豆豉及砂糖再炒約 1 分鐘，盛起。另備一蒸鑊，水滾後放入蒸 20 分鐘便成為豉椒醬。

3. 將約 2 湯匙份量的豉椒醬用匙羹略為壓爛，再加調味料與腩排撈勻，醃約 1 小時。

4. 蒸鑊水滾後放入腩排，以大火蒸約 15 分鐘，取出灑上葱花即可享用。

貼士

1. 豉椒醬材料炒過後再蒸，會令豆豉更鬆軟出味。

2. 蒜油是將蒜頭與油同炸所提煉出來的油，充滿蒜香風味，與排骨同蒸，不但增添香氣，更可令蒸出來的排骨更加香滑。

3. 腩排是靠近豬腩軟骨的部位，帶點肥肉，肉質較腍，蒸出來的口感香滑鬆軟。

自家製
脆皮燒肉

Homemade crispy
roast pork belly...202

燒臘檔的美味燒肉，大家一般都不
會想到可以在家裏自己做到，因為
總覺得這是非常複雜及麻煩。其實
並不一定喔，這個自家製燒肉，不
單製法簡易，而且效果出乎你意
料的好，咬落鬆化，皮脆肉嫩，
Juicy！還是零失敗保證成功的製
作！

材料

- 五花腩...................1 件
  - 約 3 斤

醃料

- 玫瑰露酒...............2 湯匙
- 紹興酒...................1 湯匙
- 胡椒粉...................1 茶匙
- 五香粉...................1 茶匙
- 粗鹽 A...................3 湯匙
- 白醋.......................2 湯匙
- 粗鹽 B...................350 克

工具

- 豬皮插
- 錫紙

106

做法

1. 豬腩肉洗淨，瀝乾水分，起出排骨，先將玫瑰露酒及紹興酒混合，均勻抹在腩肉上，再均勻灑上胡椒粉、五香粉及粗鹽 A，將肉反轉皮向上放入容器，在雪櫃醃一晚。

2. 隔日取出腩肉，以豬皮插在豬皮上均勻插遍整件豬皮，放在錫紙上，以錫紙摺成盤狀盛着腩肉，再在豬皮上均勻抹上白醋及鋪上粗鹽 B。

3. 預熱焗爐攝氏 200 度，將腩肉放入焗約 50 分鐘，取出，撥開及抹淨鋪在豬皮上的粗鹽，再用豬皮插均勻地在豬皮上再插一遍。

4. 將焗爐加熱至攝氏 270 度，將腩肉直接放在烤架上，再焗約 30 分鐘，直至豬皮爆滿脆泡及鬆化，取出，讓腩肉稍為冷卻，便可斬件享用了。

貼士

1. 如豬皮上仍有豬毛的話，要先用廚用火槍將毛燒掉。

2. 剛焗起的燒肉不要馬上進食，要先讓它冷卻一下，令熱氣及水分揮發後，豬皮吃起來才會香脆鬆化。

3. 以豬皮插插豬皮時，一定要整塊豬皮都要插遍，讓豬皮在烤焗過程中可以爆起香脆的脆泡，這樣豬皮吃起來才有香脆鬆化的效果。

4. 吃剩的隔夜燒肉，皮已經不脆了，不想浪費，你可配以鹹蝦醬蒸，或配以鹹蝦醬、砂糖及韮菜炒，都是非常美味的廚餘再生美味小菜。

麻香蝦子香茜伴鹹雞

Ma La salted chicken salad
with cucumber, coriander
and shrimp roe...203

材料

- 新鮮雞髀 .... 2 隻
- 花椒 .... 1 湯匙
- 粗鹽 .... 1 湯匙
- 芫茜 .... 4-5 棵
- 小黃瓜 .... 1 條
- 蝦子 .... 適量
- 炒香芝麻 .... 適量

調味料

- 麻油 .... 約 1 1/2 湯匙
- 花椒油 .... 約 3/4 湯匙

這是一道集四川麻香及廣東鹹雞風味，配以新鮮芫茜的清香及小黃瓜的爽脆，結合而成的一道餐前涼拌小菜，鹹香滋味、開胃醒神，而且製作簡易，四季皆宜。仍未想到今晚煮甚麼嗎？就來試試這道拌鹹雞吧！

做法

1. 雞髀洗淨，吸乾水分。花椒與粗鹽混合，用乾鑊以慢火炒至焦黃及散出香氣成花椒鹽，盛起，涼卻後備用。

2. 將花椒鹽均勻擦滿雞髀，用保鮮紙包好，放入雪櫃醃一晚。

3. 隔日將雞髀取出，用水沖去雞髀上的花椒鹽；蒸鍋燒滾水，放入雞髀，以大火蒸 15 分鐘至熟透，取出後倒去蒸碟排出的水分，放在通風處約 2 小時，讓雞髀充分涼卻及吹乾。

4. 芫茜及小黃瓜洗淨，瀝乾水分；芫茜切段，小黃瓜滾刀切成小塊，雞髀手撕起肉，全部放入大盆中，倒入麻油及花椒油，撈勻後上碟，灑上蝦子及炒香芝麻即可享用。

貼士

1. 雞髀洗淨後一定要吸乾水分，否則會稀釋了花椒鹽的獨特風味。

2. 蒸好的雞髀待涼卻後還要讓它自然吹乾一下，令雞髀肉乾身一些，這樣與麻油、花椒油及其他材料拌勻後，口感才不會太濕，伴以其他配料吃起來才乾爽有層次，充滿雞肉鹹香與芫茜、小黃瓜清新口味結合的風味。

3. 買回來的蝦子可以加些浸軟的陳皮絲及新鮮蒜片，用乾鑊以很慢的小火慢慢炒烘，至陳皮、蒜片變乾身後，用隔篩去掉陳皮及蒜片，涼卻後入樽保存，這樣處理過的蝦子便會更香及更加耐存。

4. 以花椒鹽醃製過的雞髀，配以廣東的臘腸、膶腸或臘肉，用來做港式煲仔飯也相當好吃。

化骨茄汁秋刀魚

Bone-softened
Pacific saury
in tomato sauce...204

「茄汁沙甸魚」可能是大家在颱風期間的最愛罐頭食品之一，通常一罐已經可以「送」幾碗飯。不過罐頭始終有添加劑，其實自家製都可以做出罐頭的化骨效果。今次我用秋刀魚取代沙甸魚，因它於秋天比較肥美當造，而且番茄的茄紅素有抗衰老的作用，實在是一款簡單健康又養顏的菜式。

「自家製」幾時都健康有益得多！

材　料

- 秋刀魚 .... 6 條
- 乾葱 .... 5 粒
　　　　　　　　　切茸
- 蒜頭 .... 5 粒
　　　　　　　　　切茸
- 指天椒 .... 1-2 隻
- 葱 .... 約 3-4 棵
　　　　　　　一開二，切段

番茄汁及調味料

- 番茄茸 .... 500 克
- 香葉（月桂葉）.... 3 片
- 魚露 .... 3 茶匙
- 砂糖 .... 6 茶匙
- 鹽 .... 1 茶匙
- 雞粉 .... 2 茶匙
- Tabasco 辣椒汁 .... 2 湯匙
- 喼汁 .... 3 湯匙
- 青檸汁 .... 1 個份量
- 黑胡椒碎 .... 適量

112

做法

1. 秋刀魚解凍後劏好，清理內臟及洗淨，吸乾水分，於中間一開為二，備用。

2. 鍋中下適量油，先爆香乾葱茸及蒜茸，下葱段在鍋底，下指天椒、番茄汁及調味料，最後放入秋刀魚，大火將汁煮滾後加蓋，以小火燜煮約2小時，至魚骨酥化，熄火，不要開蓋，至鍋內溫度自然降低後，放入雪櫃保存讓它入味，隔天翻熱後食用。

貼士

• 此做法亦可以沙甸魚代替，各俱風味。指天椒的份量可因應個人愛辣的程度而增多或減少。

椰菜花乾
牛肝菌燜醃肉

Braised pork belly with
dried cauliflower and
porcini mushrooms...205

經過曬製後的椰菜花乾，有一種比新鮮椰菜花更香濃的菜味，味道更有層次，吃起來還多了一種新鮮蔬菜所沒有的獨特煙韌口感。

椰菜花乾很適合用來燜肉，因為它在燜煮的菜式中，很能發揮到吸味的作用。除此之外，用來做涼拌菜、滾湯等效果都相當理想！大家不妨一試！

材料

- 五花腩......................2 斤
- 乾牛肝菌..................約 70 克
- 椰菜花乾..................約 70 克
- 水..........................適量

調味料

- 蠔油......................150 克
- 柱候醬....................35 克
- 磨豉醬....................35 克
- 南乳......................35 克
- 腐乳......................35 克
- 麻醬......................35 克
- 老抽......................少許
- 冰糖......................70 克

做法

1. 五花腩洗淨,用火槍燒去多餘豬毛,整塊飛水,以大火蒸 45 分鐘,取出浸冰水至冷卻後切成約 2 吋 x 2 吋方塊形狀,備用。

2. 牛肝菌略洗,浸水約半日至軟身,隔起後汁液留用。

3. 椰菜花乾略沖洗一下,輕輕飛水即撈起,備用。

4. 鍋中下油燒熱,爆香調味料,下腩肉炒勻至微微焦香,下水至差不多蓋過腩肉,以大火煮滾後燜約 10 分鐘後改小火燜約 45 分鐘,下牛肝菌、椰菜花乾及浸過牛肝菌的水,再燜 5 分鐘,熄火再焗 15 分鐘即成。

貼士

1. 另一吃法是在腩肉改小火燜約 30 分鐘後,以大火將汁收「傑」,這樣腩肉及汁液的味道會較為濃稠,另有一番滋味。

2. 除了用腩肉之外,還可以腩排代替,各俱風味。

3. 腩肉先蒸過後才燜煮,會使腩肉的形狀保持得較為完整,令賣相更加美觀。

煮完豬腩肉，
洗乾淨個煲，
再煮下一味餸，
狠好玩呀！

仁稔醬蒸排骨

Steamed pork ribs
in Renren sauce...20g

材料

- 腩排......................1 斤
  斬件
- 仁稔醬......................200 克
- 葱花......................適量

調味料

- 砂糖......................1 茶匙
- 雞粉......................1 茶匙
- 鹽......................1/2 茶匙
- 油......................1 茶匙

做法

1. 排骨洗淨瀝乾水分,與調味料撈勻醃約 30 分鐘,之後再加入仁稔醬撈勻放入碟中,備用。

2. 蒸鑊煲滾水,放入排骨,加蓋,以大火蒸約 8 分鐘,取出,灑上適量葱花即可享用。

貼 士

- 排骨要預先用調味料醃約30分鐘,讓排骨先有一個「底味」,然後才加入仁稔醬,這樣蒸出來的排骨才會夠入味,而仁稔醬味道亦會更鮮明。

仁稔醬配肉類燜、蒸、煮,都各有風味,它那微酸的味道,能夠中和肉類的肥膩,亦令人胃口大開,是夏日煮餸的最佳配料。

蜜餞脆皮欖角豆豉雞

Crispy chicken in black bean and olive honey glaze...206

材 料

- 全雞 .... 1 隻
  約 2 1/2 斤
- 欖角 .... 40 克
- 豆豉 .... 8 克
- 蜂蜜 .... 120 克
- 生粉 .... 適量

調 味 料

- 鹽 .... 6 克
- 砂糖 .... 4 克
- 雞粉 .... 4 克
- 生粉 .... 20 克
- 水 .... 共 4 湯匙
  分兩次用

雞的烹調及吃法實在太多了，我們嘗試一下甜味道的蜜餞口味吧！醬汁除了蜂蜜之外，還加入了鹹鹹香香、風味獨特的廣東經典食材—欖角，鹹甜之間互相調和，香脆惹味，絕對是舌尖上又一新的美味！

做法

1. 雞洗淨，先起出腿骨、胸骨及背骨，斬件；欖角及豆豉洗淨，吸乾水分，剁碎，備用。

2. 調味料先用 2 湯匙水撈勻，再與雞件撈勻醃約 1 小時，備用。

3. 雞件平均撲上適量生粉；於平底鑊下適量油，燒熱，下雞件（雞皮向下），以中火將雞件煎至兩面金黃香脆，盛起，瀝乾油分，備用。

4. 鑊中再下適量油，爆香欖角及豆豉，下水 2 湯匙炒勻，下蜂蜜再炒勻，煮至起泡，熄火，馬上將雞件回鑊，與醬汁撈勻後即可上碟享用了。

貼士

1. 斬雞前先起出腿骨、胸骨和背骨，可更易於進食，減少咬到骨頭的機會，亦可令雞件更加香脆。

2. 煎雞前才撲上生粉，若太早撲上生粉的話，雞件的水分會稀釋及溶解生粉，大大影響了雞件煎出來的香脆效果。同時，將雞皮向下煎，除可使雞皮的油分滲出之外，亦可令雞皮煎得更加香脆。

3. 當蜜餞醬汁煮至起泡後，便要熄火，還要馬上將雞件放入撈勻，原因是此蜜餞醬汁主要是蜂蜜，沒有生粉埋芡的濃稠度（蜂蜜亦不適宜高溫長時間烹煮），若繼續煮的話，它仍呈液態，如雞件放入煮得太久，便會大大影響了雞件的香脆度。

4. 此做法除了用雞之外，我試過用豬腩排效果也相當不錯，亦有異曲同工之妙，各有風味。

鮮沙薑蜆蚧脆皮燒雞

Crispy roast chicken stuffed with preserved clams and sand ginger ...207

雞的吃法與烹調方法，真的可以說是千變萬化，在不同的地方，不同廚師的手裏，用雞做出來的菜式，可以說是各自各精彩，各有特色，並深受食客歡迎！

這道燒雞菜式，融入了帶有東南亞風味的鮮沙薑，配以潮洲菜的蜆蚧醬，鹹香惹味中夾雜着鮮沙薑的辛辣濃郁的香氣，非常滋味獨特。而且用了非常討好的烤焗烹調法處理，皮脆肉嫩，相信大家一定會囍歡！

材料

- 雞 .... 1 隻
- 鮮沙薑 .... 40 克
- 蜆蚧 .... 35 克
- 肉碎 .... 75 克
- 生抽 .... 1 茶匙
- 老抽 .... 少許
- 生粉 .... 1/2 茶匙
- 麻油 .... 1 茶匙
- 蠔油 .... 1 茶匙
- 糖 .... 1 茶匙
- 水 .... 60 毫升

醃料

- 粗鹽 .... 20 克
- 沙薑粉 .... 20 克

沙薑醬

- 鮮沙薑 .... 80 克
- 葱 .... 25 克
- 橄欖油 .... 120 克
- 沙薑粉 .... 2 克
- 雞粉 .... 2 克
- 幼鹽 .... 2 克
- 白醋 .... 3 克

做法

1. 沙薑醬：沙薑及葱洗淨，完全瀝乾水分及吹乾。沙薑切碎，葱切成細葱花，與其他材料撈勻即可。

2. 雞洗淨去內臟，抹乾後用粗鹽、沙薑粉抹勻全身內外，再用保鮮紙包好放入雪櫃醃一晚，備用。

3. 肉碎加入生抽、生粉，醃 10 分鐘。

4. 鮮沙薑洗淨，切碎，備用。

5. 燒熱油鑊加入蜆蚧炒香，撈起，下少許油再將鮮沙薑炒香，隔起備用。

6. 將肉碎炒熟後，加入已炒香的蜆蚧、沙薑，加水，並下麻油、蠔油及糖調味，再用生粉打芡，加少許老抽，做成沙薑蜆蚧肉餡。將肉餡放入雞肚內，用鵝尾針縫好，再用保鮮紙包好放入雪櫃醃一晚。

7. 焗爐預熱攝氏 150 度，將雞入爐焗 60 分鐘後，將溫度升至攝氏 250 再焗 15 至 20 分鐘，至雞皮金黃香脆，即可享用。

美味家常茄汁燜牛肉

Home-style braised beef shin in tomato sauce...208

最受主婦歡迎的家常菜，一定要製作簡易而又要做出來好吃，還要「好餸飯」，這道茄汁燜牛肉，不單烹煮非常容易，而且酸辣可口，相當開胃好吃，還可一次過做好一大鍋，吃時翻熱便可，確是一道很適合家庭製作的美味小菜。

此菜除了可配飯之外，配以湯麵也相當好吃，只要將湯麵煮好之後，將牛肉連茄汁淋在湯麵上，加些蔥花或芫茜，便成為一碗美味醒胃的牛肉麵了。

材料

- 新鮮牛腱 .... 1.6 公斤
- 洋蔥 .... 1 個
- 薯仔 .... 4 個
  約 900 克

茄汁及調味料

- 番茄茸 .... 1 公斤
- 雞湯 .... 500 毫升
- 香葉（月桂葉）.... 6 片
- 指天椒 .... 3-4 隻
- 魚露 .... 6 茶匙
- 砂糖 .... 60 克
- 鹽 .... 1 茶匙
- 雞粉 .... 4 茶匙
- Tabasco 辣椒汁 .... 2 湯匙
- 喼汁 .... 3 湯匙
- 青檸汁 .... 1 個份量
- 黑胡椒碎 .... 適量

做法

1. 牛腱洗淨，切件，飛水。洋葱去皮，洗淨，切粒。薯仔洗淨，去皮，切件，用清水泡浸着，以防氧化變黑，備用。

2. 鍋中下適量油，先爆香洋葱，將牛腱、茄汁及調味料下鍋大火煮滾後，改以小火燜煮約 1 小時 30 分鐘，下薯仔再繼續燜約 30 分鐘，至牛腱及薯仔燜至鬆軟及腍，熄火，即可上碟享用了。

貼士

1. 此菜可完成後不用開蓋，讓牛肉及薯仔利用鍋內的餘溫再焗煮，待涼卻後放入雪櫃，隔天翻熱食用會更入味、風味更佳。

2. 薯仔一般燜約半小時便腍，所以薯仔一定要在最後半小時才下，薯仔若燜得太久的話，便會太爛，失去薯仔吸收湯汁後的美味質感。

我的珍藏

從小，我便有收藏物件的興趣及習慣，小學到中學生時代，受當時的風氣及同學影響，便跟著大家一起玩當時學生最流行的集郵，儲下了很多很多郵票。出來社會工作之後，開始收集的東西越來越多，火柴盒、唱片、影碟、畫作、陶瓷、玻璃藝術品等等。當然還有世界各地不同類型的食譜，從我開始愛上烹飪以來，我收藏的食譜起碼超過幾千本，當然，有些純粹是因為覺得食譜製作精美，圖片吸引而購買的。

而一直以來，最令我瘋狂收藏的就是不同風格類型的各種餐具、廚具及咖啡杯。

無論公幹或旅遊，除了當地的街市、超市外，我一定還會逛逛各地的家品店及百貨公司的廚具部。我很喜歡有當地特色，設計美觀而又實用的各種餐具、廚具。除了用來收藏之外，它們還有很多實際用途，例如拍攝食譜、訪問、烹飪示範又或是在家請客，我都會在這些收藏中揀選適合的、喜歡的派上用場。

但我亦相信很多「煮人」亦會跟我一樣，有些特別囍歡的餐具會被列入珍藏類，意思就是只會長期用作擺設，或長期放在櫃子裏，不輕易拿來使用，只會偶爾拿出來觀賞或請些特別 VIP 飯聚時，才小心翼翼地拿出來「使用」及炫耀一下。

現在，無論在我家及餐廳的櫃子裏，所有可以置放東西的空間，都被我這些大大小小的廚具、餐具佔領了。每次，當我打開櫃門，看到這些各式各樣的餐具、廚具，都有一種令我覺得幸福的、滿足的開心感覺。

黑蒜蒜茸包
Black garlic bread...210

材料

- 淨肉黑蒜 .... 150 克
- 新鮮蒜茸 .... 150 克
- 幼鹽 .... 約 1 1/2 茶匙
- 日本牛油 .... 250 克
- 法國麵包或任何自己喜愛的麵包，切成適當厚度

自從在專欄教了大家製作零失敗電飯煲焗黑蒜之後，反應非常熱烈，還引起了很大的迴響，很多人都成功地在家輕易製成了完美有益的黑蒜。

之後無論在烹飪班、Facebook，甚至是在街上，很多人都走來問我，黑蒜除了就這樣吃之外，還可以有甚麼不同的吃法？可以變化出甚麼菜式呢？好吧！我就跟大家分享這道零失敗簡單易做，而且美味健康的黑蒜菜式！

做法

1. 先將黑蒜肉用叉壓爛成黑蒜茸，再與新鮮蒜茸、幼鹽及牛油混合，然後攪拌均勻，令全部材料完全融合，成黑蒜蒜茸醬，然後裝入可密封的容器並且蓋好，放入雪櫃冷凍定形及保存。

2. 吃時將適量黑蒜蒜茸醬抹在麵包上，預熱焗爐約攝氏 180 度，將麵包放入焗 3 至 4 分鐘（視乎麵包厚薄或種類），至麵包微焦帶脆，便可取出享用了。

貼士

1. 我建議選用日本出品如雪印、明治等牌子的牛油，因為這些日本出產的牛油質地比較柔軟，不但在製作時更易掌握控制，焗出來也特別香滑好吃。

2. 在剁蒜茸前，蒜頭一定要徹底抹乾水分，不可有任何生水，這樣做出來的黑蒜蒜茸醬便可保存更耐，不易變壞了。

3. 此黑蒜蒜茸醬的材料比例，都是根據我自己的口味而設定，大家可以此比例做一次，然後再根據自己的口味，在黑蒜、新鮮蒜茸及牛油的比例上，加多或減少。

黑松露蝦多士

Black truffle shrimp toast...211

「蝦多士」是一道很港式的地道小食，在我小時候的茶餐廳及酒樓都有此菜式供應，鮮味香口，很受歡迎。此菜製作亦不難，而且很容易掌握及做得好，我今次還在蝦膠內加入了蟹籽，令蝦膠咬起來多了一種脆卜卜的快感，同時亦加入了現在中菜很流行用的冰鮮黑松露，令這懷舊菜式除多了一份矜貴之外，還增添了一股迷人的香氣。

材料

- 方包 .... 4 塊
- 鮮蝦仁 .... 12 隻
- 蝦膠 .... 180 克
- 蟹籽 .... 15 克

調味料

- 鹽、砂糖、雞粉 各少許
- 胡椒粉、生粉 ..... 各少許

黑松露醬

- 冰鮮黑松露 .... 1 粒
  約 30 克
- 橄欖油 .... 約 3 湯匙
- 鹽 .... 少許

做法

1. 方包切去麵包皮，再大約一開三切開成小塊。蝦膠與蟹籽、調味料撈勻；蝦仁開邊、挑腸，用少許鹽撈勻略醃，備用。

2. 黑松露略洗，吸乾水分，剁碎，與橄欖油及鹽撈勻成松露醬，備用。

3. 將處理好的蝦膠平均抹於小方包上，再放上蝦仁。鑊中燒熱油，以中火將蝦多士炸至金黃香脆，撈起，用廚紙吸淨油分，上碟，再在蝦仁上放適量黑松露醬及裝飾便可享用。

貼士

1. 如想方便快捷，蝦膠可以買現成的。若自己做亦不難，只要將蝦仁挑腸，用刀拍爛，剁碎，加入調味料然後再撻至起膠即成。

2. 處理油炸食物時，在食物快要炸好起鑊前，將火加大再炸約十數秒才將食物撈起，這樣可使食物內的水分逼出，令食物更香脆。

3. 此蝦多士很適合派對，可大批製作，就算做好後放一至兩小時後仍很香脆可口。

4. 冰鮮黑松露可在一些專賣菇菌的專門店購買。

冬蔭功海皇天使麵

Seafood angel hair pasta in Tom Yum sauce...214

材料

- 天使麵..............200 克
- 花蛤..............6 隻洗淨
- 鮮中蝦..............6 隻洗淨、去殼、挑腸，蝦頭、蝦殼留用
- 帶子..............6 隻洗淨、吸乾水
- 鮮魷魚..............約 50 克洗淨、切片
- 泰國小番茄..........5 粒一開二
- 椰汁..............約 1/4 杯

冬蔭功汁

- 乾葱頭..............15 克略切
- 蒜頭..............15 克拍鬆
- 香茅..............2 枝只要根莖部分、切圈
- 檸檬葉..............8 片用手搓裂
- 南薑..............70 克切片
- 指天椒..............2 隻切圈
- 冬蔭功醬..........60 克
- 水..............600 毫升

調味料

- 鮮榨青檸汁........約 3 個份量
- 魚露..............25 克
- 砂糖..............15 克

或許是受童話故事影響，一直認為天使是穿着一身潔白長袍、有一對可愛的翅膀、頂着一頭柔軟的頭髮、並擁有溫柔笑容的天國使者。所以當聽到有一款意大利麵被稱為「天使麵」（Angel hair）時，感覺命名者還真是個浪漫的人呢！

今次特別選用天使麵，因其柔軟細膩的麵質更能吸收冬蔭功的湯汁。當火辣誘人的冬蔭功與纖細的天使麵結合時，真的會令人上癮啊！

做法

1. 先將乾葱頭、蒜頭、蝦頭及蝦殼炒香，並用鑊剷將蝦頭壓爛，令蝦膏溢出，加水，加入所有冬蔭功汁料，煮約 30 分鐘至湯汁濃郁至約 350 毫升，下調味料，試味可以後隔起湯渣，備用。

2. 天使麵以滾水焓約 3 分鐘，撈起，瀝乾水分，備用。

3. 海鮮用滾水灼或炒至約五成熟，備用。

4. 鑊中加入冬蔭功汁煮滾，加入天使麵煮約 2 至 3 分鐘，至湯汁開始收乾時加入海鮮及番茄炒勻，加入椰汁繼續煮至海鮮全熟及湯汁收乾，即可盛起上碟享用了。

貼士

1. 在製作做法（3）時，海鮮煮至約五成熟即可撈起，不要全熟，否則在海鮮回鑊與天使麵一起煮時，海鮮便會過熟而變「老」，大大影響了海鮮原有的鮮嫩質感。

2. 在製作做法（4）時，亦可因應個人喜愛吃意粉的軟硬程度，在湯汁份量及烹煮時間上加多或減少。

3. 下調味料時，亦可因應個人的濃淡口味而增多或減少。

自家製
泡菜金蠔Pizza

Kimchi and semi-dried oyster pizza...212

公司附近有一間麵包工場，每天早上都飄來陣陣的麵包香氣，就算已經用過早餐，我的胃還是很不爭氣的嚷着「好香，好想吃！」。麵包機近年大行其道，為我這個生活忙碌的麵包愛好者帶來很大的便利。

後來慢慢學習不依賴麵包機，發現全部由自己親手製作，原來樂趣更大！像這款「泡菜金蠔Pizza」是我的延伸之作，由搓粉糰至烤焗完成都只是簡單幾個步驟。其實配料可按個人口味自由配搭的，只要加上創意，你就可以做出一個充滿個人風格的自家製pizza了。

Pizza 餅底材料

- 高筋麵粉 .... 300 克
- 橄欖油 .... 20 克
- 砂糖 .... 1/2 茶匙
- 鹽 .... 5 克
- 速發酵母 .... 2.5 克
  先用少許溫水調勻
- 溫水 .... 200 克

Pizza 餡料

- 韓國泡菜 .... 500 克
  揸乾汁液，切成適當大小
- 意粉茄醬 .... 約 100 克
- Mozzarella 芝士 .... 200 克
- 金蠔 .... 6-8 隻視乎大小，用水浸軟，洗淨，摳去頂部的
  「枕」，吸乾水分，切成適當大小
- 柚子蜜 .... 約 80 克
- 紹興酒 .... 少許

做法

1. 將高筋麵粉、橄欖油、砂糖、鹽先撈勻,然後下酵母再撈勻,再慢慢加入餘下的溫水,邊加邊搓,直至成粉糰後,再將粉糰搓至表面光滑。(如用機打,可先用慢速,到搓成粉糰後再以中速打至粉糰表面光滑。)

2. 將搓至光滑的粉糰放入容器,蓋上保鮮紙,放在室溫發酵約 1 小時,至粉糰膨脹及帶有筋性。(如在天氣乾燥的季節時製作,可在粉糰搓至光滑後,在粉糰上噴灑少許水,然後再用一張濕的廚紙蓋着粉糰,以確保粉糰濕潤。)

3. 將發酵完成的粉糰取出,在枱面上灑上適量麵粉,再將粉糰用麵粉棍推壓成適當大小及厚薄的 Pizza 餅底。

4. 焗盤灑上麵粉,放上餅底,略為整理及修整餅邊,並用叉在餅底上均勻地「篤」上叉孔,以防止餅底在烤焗時有氣泡膨脹,令餅底凹凸不平。預熱焗爐攝氏 180 度,將餅底放入,焗約 10 分鐘成 Pizza 餅底。

5. 在烤焗餅底期間,將鑊燒熱,倒入柚子蜜及紹興酒,攪勻,煮熱至起泡,放入金蠔撈勻,盛起,備用。

6. 餅底焗好後取出,先在餅底上搽上茄醬,再依次均勻放上芝士、泡菜及金蠔,再放上芝士。再預熱焗爐攝氏 220 度,放入 Pizza,焗約 10 分鐘至芝士溶化及餅邊金黃香脆,便可取出 Pizza,要趁熱享用。

貼士

1. 酵母先用水調勻,在搓粉糰時可更易於與粉糰均勻融合。

2. 在做法(1),酵母不可同時與鹽直接放在一起,一定要先將高筋麵粉、橄欖油、砂糖、鹽先撈勻,才下酵母,因為鹽會殺死酵母,影響酵母的活躍性。

3. 在潮濕的環境下,酵母會較為活躍,因此,搓好的粉糰不宜放在雪櫃發酵,一定要放在室溫,同時在不同潮濕度及乾燥的季節,發酵時間會因此而可能有所加長或減短。

4. 泡菜在放入餅底前一定要先揸乾水分,否則太濕的話,會大大影響餅底的鬆脆度。

仁穩醬薑蔥蝦子撈麵

*Egg noodles dressed
in Renren sauce, garlic scallion oil
and shrimp roe...215*

材料

- 全蛋麵 .... 2 個
- 葱蒜油 .... 約 1 湯匙
- 蝦子 .... 約 1 湯匙
- 仁稔醬 .... 2-3 湯匙
- 葱花 .... 適量

葱蒜油

- 葱 .... 4 棵
  切段
- 蒜頭 .... 6 粒
  拍鬆
- 油 .... 1/2 杯

做法

1. 葱蒜油做法：鑊中燒熱油，放入葱及蒜頭，以慢火炸煮約 10 至 15 分鐘，至葱及蒜頭變焦黃及散發出香氣，隔去蒜頭及葱後，成葱蒜油。

2. 全蛋麵用滾水煮至軟身，過冷河，再煮熱，撈起，瀝乾水分，上碟；淋上葱蒜油，灑上蝦子，加上仁稔醬，最後灑上葱花，吃時將全部材料撈勻享用。

貼士

1. 此撈麵除了以全蛋麵製作之外，基本上用任何你自己喜愛的各種麵類或米粉製作均可，各俱風味。

2. 將全蛋麵煮軟後過冷河再煮熱，可令麵條吃起來更爽口。

3. 仁稔醬在使用前要用微波爐稍為加熱，這樣仁稔醬才夠軟身，更易與麵條撈勻，以及可令仁稔醬內的油分溶化，吃起來更香、更可口。

辣酒公仔麵煮蜆

Clams in spicy sauce with instant noodles...218

材料

- 蜆......................約 1 1/2 斤
- 任何即食麵..........約 1 包
- 乾葱茸、蒜茸.....各 30 克
- 指天椒.................約 5 隻
  切圈
- 芫茜.....................約 3-4 棵
  洗淨，切段，留適量切碎
- 雞湯.....................1 杯
- 玫瑰露酒.............25 毫升
- 紹酒.....................15 毫升
- 米酒.....................15 毫升
- 油........................約 80 毫升

調味料

沙爹醬....................50 克

鹽..........................1/3 茶匙

砂糖.......................1/2 茶匙

香辣菜式在配搭及烹調上真是變化多端，可塑性非常之高，這道「辣酒公仔麵煮蜆」均集合了辣椒、乾葱、蒜、沙爹醬的各種不同辛辣，香辣元素，再配以玫瑰露撲鼻的酒香，中國經典廚酒紹興酒的甜香酒味，以及帶有辛辣酒香的廣東米酒做成的濃郁湯汁，令花蛤增添了惹味的辣味之外，還帶着香噴噴的醉意。因為此湯汁實在惹味，我特意加入了「索」味的即食麵，可以將精華盡吸其中，美味非常！

做法

1. 蜆先浸水 2-3 小時，待吐沙後用刷子擦洗乾淨，瀝乾水分，備用。

2. 鑊內燒熱油，先下乾葱茸炒香，再下蒜茸及指天椒炒至散發出香氣及蒜茸開始變金黃色，下沙爹醬炒勻，下蜆炒勻，下雞湯、玫瑰露酒、紹酒及米酒煮開，下鹽、糖調味，煮至蜆完全張開至全熟，下芫茜炒勻；與此同時，另備一鍋煲滾水，下即食麵煮熟，用篩隔起水分，將即食麵放於盛器底，然後再將剛才煮好的蜆連同湯汁倒入盛器中，最後灑上芫茜碎即成。

貼士

1. 此道菜式之蜆可以花蛤、瀨尿蝦、花螺、鮮蝦或蟹等其他海產代替；即食麵部分更可以粉絲、烏冬、葛絲、米粉甚至油條等澱粉類食材代替，以吸收此惹味湯汁。

2. 我還試過以此湯汁的份量加大配以蘿蔔、冬菇、香芹及豆腐變成火鍋湯底，惹味非常，在即將來臨的秋冬火鍋季節，大家不妨一試。

3. 指天椒的份量可以因應個人愛辣的程度而增多或減少。

惹味冬蔭功 Pizza

Tom Yum seafood pizza...216

冬蔭功是泰國菜的經典代表菜式之一，它酸辣鮮香的獨特味道真是非常吸引，深受香港人的歡迎，亦是到泰國菜館必點的名菜。

在家也可以輕鬆炮製出充滿泰國風味的冬蔭功 Pizza，效果保證大家喜出望外，因為這 Pizza 實在「泰」好食！「泰」滋味了！

Pizza 餅底材料

- 高筋麵粉 .... 300 克
- 橄欖油 .... 20 克
- 砂糖 .... 1/2 茶匙
- 鹽 .... 5 克
- 速發酵母 .... 2.5 克
  先用少許溫水調勻
- 溫水 .... 200 克

Pizza 餡料

- 鮮魷魚 .... 150 克
  切圈或切條
- 鮮中蝦 .... 約 10 數隻
  去殼、挑腸
- 北海道帶子皇 ..... 約 4 隻
  一開四
- 泰式冬蔭功醬 ..... 適量視乎愛辣程度
- 意粉茄醬 .... 約 100 克
- Mozzarella 芝士 .. 200 克
- 金不換 .... 1 棵
  只要葉部分

調味料

- 鹽 .... 少許

做法

1. 將高筋麵粉、橄欖油、砂糖、鹽先撈勻，然後下酵母再撈勻，再慢慢加入餘下的溫水，邊加邊搓，直至成粉糰後，再將粉糰搓至表面光滑。（如用機打，可先用慢速，到搓成粉糰後再以中速打至粉糰表面光滑。）

2. 將搓至光滑的粉糰放入容器，蓋上保鮮紙，放在室溫發酵約 1 小時，至粉糰膨脹及帶有筋性。（如在天氣乾燥的季節時製作，可在粉糰搓至光滑後，在粉糰上噴灑少許水，然後再用一張濕的廚紙蓋着粉糰，以確保粉糰濕潤。）

3. 約 1 小時後，取出發酵完成的粉糰，在枱面上灑適量麵粉，再將粉糰用麵粉棍推壓成適當大小及厚薄的 Pizza 餅底。

4. 焗盤灑上麵粉，放上餅底，略為整理及修整餅邊，並用叉在餅底上均勻地「篤」上叉孔，以防止餅底在烤焗時有氣泡膨脹，令餅底凹凸不平。預熱焗爐攝氏 180 度，放入餅底，焗約 10 鐘成 Pizza 餅底。

5. 在烤焗餅底期間，將魷魚、蝦及帶子吸乾水分，下適量鹽撈勻略醃，然後炒至半熟，盛起，瀝乾水分，備用。

6. 餅底焗好後取出，先在餅底上均勻搽上冬蔭功醬及茄醬，再在餅底上均勻放上魷魚、蝦及帶子，最後均勻灑上芝士。再預熱焗爐攝氏 220 度，放入 Pizza，焗約 10 分鐘，至芝士溶化及餅邊金黃香脆，便可取出 Pizza，灑上金不換葉後，便可趁熱享用了。

貼士

1. 酵母先用水調勻，在搓粉糰時可更易於與粉糰均勻融合。

2. 在做法（1），酵母不可同時與鹽直接放在一起，一定要先將高筋麵粉、橄欖油、砂糖、鹽先撈勻，才下酵母，因為鹽會殺死酵母，影響酵母的活躍性。

3. 在潮濕的環境下，酵母會較為活躍，因此，搓好的粉糰不宜放在雪櫃發酵，一定要放在室溫，同時在不同潮濕度及乾燥的季節，發酵時間會因此有所加長或減短些。

4. 海鮮先用鹽略醃及炒至半熟，除了是在焗 Pizza 時減短海鮮的烹煮時間外，主要是將海鮮多餘的水分排出，這樣焗出來的 Pizza 餅底才不會太濕，保持香脆。

5. 冬蔭功醬是煮泰式冬蔭功湯的主要調味醬料，除可煮冬蔭功湯之外，還可以用來醃各種肉類、炒飯及各類菜式的調味，可於一般泰國食品雜貨店買到。

南瓜海皇冬蔭功

Tom Yum Goong
pumpkin seafood soup...220

「冬蔭功蝦湯」，是泰國菜之中非常經典及受歡迎的其中一道，鮮、香、酸、辣、鹹、甜共冶一鍋，我亦十分喜歡這款湯，因為非常醒胃、惹味。這款冬陰功，我特意加入了帶有甜味的南瓜，因為我覺得南瓜的獨特甜味在泰式酸辣味道中非常突出及合襯，尤其將南瓜煮至軟爛與湯融合的時候，可令湯底更加濃稠及香甜。另外，除了蝦之外，我還加入了蟹及花蛤，令此湯倍添鮮美！

材料

- 鮮中蝦..............10 隻
  剝殼，挑腸；蝦頭及蝦殼留用
- 肉蟹或膏蟹........1 隻
  劏好，斬件
- 花蛤..............10 隻
  洗淨，浸水
- 蒜頭..............5 粒
  切片
- 乾葱頭............5 粒
  切碎
- 香茅..............8 枝
  只要根莖部，切碎
- 南薑..............40 克
  切片
- 芫茜..............約 6 棵
  切碎，芫茜頭留用
- 指天椒............2-4 隻
  視乎愛辣程度，切碎
- 檸檬葉............10 克
  洗淨，用手搓裂
- 南瓜..............450 克
  去皮、去核、切塊
- 泰國小番茄......250 克
  洗淨，一開二
- 草菇..............約 10 數粒
  洗淨，一開二
- 雞湯..............2 公升

調味料

- 冬蔭功醬..........約 150 克
- 魚露..............20 克
- 砂糖..............10 克
- 鮮榨青檸汁......約 4 個份量
- 椰汁..............適量（最後下）

做法

1. 鍋中燒熱油約 1/2 杯,先下蒜頭及乾葱頭炒香,再下蝦頭及蝦殼爆香,並將蝦頭壓爛,讓蝦膏溢出,下雞湯、香茅、南薑、芫茜頭、指天椒及檸檬葉,大火煮滾後加蓋,以小火煮 30 分鐘成湯底。

2. 以隔篩隔去湯渣,將湯倒回鍋內,下南瓜煮約 10 分鐘至南瓜開始變腍,下番茄、草菇及調味料拌勻,試味,煮至南瓜開始軟爛,番茄變軟,先下蟹及花蛤略煮,再下蝦灼熟後馬上淋上適量椰汁及灑上芫茜碎,拌勻即可盛起享用了。

貼士

1. 冬蔭功醬(見左圖)是混合香茅、南薑、辣椒、花生、蝦米、蒜頭、乾葱頭、羅望子汁等攪碎用油炒成的現成醬料,可在一般泰國食品雜貨店買到。由於每個牌子的酸辣鹹甜程度都有差別,建議若首次使用的話可先買幾款試味作比較,然後選一款最適合自己口味的使用。

2. 冬蔭功這道泰式經典蝦湯,靈魂所在就是浮在湯面上的那層蝦油,因此一般煮泰式冬蔭功都是使用蝦膏較多的大頭蝦,但大頭蝦不易在香港一般街市買到,因此,選用新鮮中蝦亦可,但在做法 (1) 炒蝦時就一定要炒得夠香,令蝦膏全部溢出,這樣煮出來的冬蔭功才夠鮮、香及惹味。

3. 此冬蔭功湯底亦可將它變成火鍋湯底,配以各類海鮮、肉類及各種粉、麵澱粉類食物都非常美味。

養生黑蒜
桂圓鴛鴦豆燉乳鴿

Double-steamed squab soup
with black garlic, peanuts and
black-eyed beans...209

材料

- 黑蒜.................25 克
  約半個
- 桂圓肉...............60 克
- 花生.................80 克
- 眉豆.................40 克
- 乳鴿.................1 隻
  約半斤
- 水...................1.8 公升

調味料

- 鹽...................約 15 克
- 砂糖.................約 6 克

做法

1. 黑蒜去外衣，桂圓肉、花生、眉豆洗淨。乳鴿洗淨飛水，備用。

2. 將所有材料放入容器，蓋上保鮮紙，水滾後以慢火隔水燉約四小時，隔走浮起的油分，下調味料，調至適合自己的口味後即可上枱享用了。

貼士

1. 做燉湯時一定要在容器上蓋上保鮮紙，以防蒸氣及「倒汗水」滴回容器內，影響烹調效果，同時亦可保持燉湯食材原汁原味的營養價值。

2. 一般家庭蒸爐及蒸鑊的水容量，都一定不足以蒸燉四小時，所以在蒸燉中途記得加水，以免蒸乾水分發生危險。

有人說一頓圓滿晚飯，必須要有甜品！說的不錯，飯後吃一些甜點，能叫人心情變得歡愉。不過湯水在中國人的飯桌上，卻有着比甜品更重要的地位。無論天氣多冷或多熱，喝一口熱騰騰的湯，整個人都感覺舒暢極了！這個燉乳鴿湯，特別加入黑蒜、眉豆、花生，令它更為養生有益。

黑松露野菌忌廉湯

Black truffle soup with mushrooms...219

近年自從雲南進口的冰鮮黑松露打入本港市場後，以黑松露作賣點的中西菜式便大行其道，因為價格便宜（約是法國、意大利售價的十分之一），因此各大小餐廳的菜式中，很多都加入了黑松露的元素。這道原本是名貴食材，在餐廳才可喝到的黑松露忌廉湯，現在就算在家也可以低成本自己製作，而且還是保證零失敗！

材料

- 冰鮮黑松露 .... 80 克
- 鮮忌廉 .... 150 毫升
- 鮮奶 .... 150 毫升
- 雞湯 .... 150 毫升
- 牛油 .... 10 克
- 蒜茸 .... 少許
- 日本本菇 .... 適量
- 番茜碎 .... 適量（裝飾用）

麵撈材料

- 牛油 .... 20 克
- 麵粉 .... 20 克

調味料

- 鹽、砂糖、雞粉 各少許
- 松露油 .... 適量

做法

1. 細火將鍋燒熱，下麵撈材料之牛油煮溶，下麵粉撈勻至糊狀成麵撈，盛起，備用。

2. 黑松露解凍後略洗，用廚紙吸乾水分，部分切片留用，其餘略為切碎後放入攪拌機，加入雞湯，將黑松露打至幼滑成黑松露雞湯，備用。

3. 鍋中加入鮮忌廉及鮮奶煮滾，再加入打好的黑松露雞湯，以慢火煮滾，下鹽、砂糖及雞粉調味，試味滿意後下麵撈，像埋芡一樣邊煮邊攪，煮至自己喜愛及適合的濃稠度，倒入盛器。

4. 鑊中下材料內的牛油煮溶，下蒜茸炒香，下本菇炒軟後放上湯面，淋上適量松露油，最後灑上番茜碎裝飾即可趁熱享用。

貼士

1. 此湯除了使用本菇外，亦可隨個人喜愛選擇其他的菇菌類代替，各俱風味。

2. 用攪拌機打黑松露雞湯時，要盡量打得幼滑些，這樣做出來的忌廉湯口感才會細滑，否則感覺有渣及纖維就不好喝了。

3. 冰鮮黑松露可在一些賣菇菌的專門店買到。最後下的松露油，目的是增加忌廉湯的松露香氣，松露油可於各大型日式超市有售。此食譜份量約可做出 4-6 碗。

冰火熔岩朱古力蛋糕

Warm black truffle chocolate fondant a la mode...222

材料

- 無糖黑朱古力 .... 80 克
- 牛奶朱古力 .... 80 克
- 無鹽牛油 .... 90 克
- 砂糖 .... 50 克
- 全蛋 .... 4 個
- 蛋黃 .... 4 個
- 低筋麵粉 .... 60 克
- 黑松露 .... 約 20 克
- 松露油 .... 20 克
- 牛奶或雲呢嗱雪糕 ...... 適量

這是一道改良自經典甜品「心太軟」的變奏之作，其原理其實就是一個未焗熟的蛋糕。容器周邊的蛋糕已熟，而中間的蛋糕漿仍然未熟，呈流質狀態，因此吃起來便造成一種「流心」效果，再特別加入新元素黑松露的獨特香氣，令朱古力的味道更富有層次及有新鮮感。

製作難度並不高，但效果相當有驚喜，很適合派對及宴客，保證能讓你的賓客回味無窮。

做法

1. 先將黑朱古力、牛奶朱古力及牛油放入容器內，放在熱水上慢慢攪拌至溶解；麵粉用粉篩篩過，黑松露切碎，備用。

2. 將全蛋及蛋黃加入砂糖打勻成蛋漿（只需全部打勻便可以，切勿將蛋打得太久至打起狀態，否則焗出來的蛋糕會脹大，不是此甜品所需要的效果）。

3. 將已溶化的朱古力及牛油加入蛋漿內打勻，再加入麵粉、黑松露碎及松露油攪勻至粉漿幼滑無粉粒，最後倒入適合焗爐用的容器內約七至八分滿。

4. 預熱焗爐攝氏 200 度，放入焗約 10 分鐘，以手指輕按靠近容器旁邊的蛋糕已經變熟凝固，再以長竹籤插入距離容器約 1.5 厘米以內的位置，若抽出來的竹籤仍是流質狀態，則表示這個蛋糕已經成功了。

5. 將蛋糕取出，在蛋糕上放上雪糕球及裝飾伴食即可享用。

貼士

1. 黑松露要盡量切得碎些，這樣整個蛋糕吃起來的口感便會細滑些。

2. 不要使用高筋度的麵粉來做此甜品，因為這樣會令焗出來的蛋糕質感帶有筋性，不夠鬆軟；同時麵粉亦要用篩篩過，這樣會令麵粉更幼細，焗出來的蛋糕亦會更鬆軟有彈性。

3. 各家焗爐的火力及溫度會各有參差，擺放入焗爐的蛋糕數量均會影響焗出來蛋糕的生熟及軟硬度，所以取出時一定要測試一下，以達至這蛋糕最理想的效果。

4. 配以此蛋糕的雪糕最理想及適合的，我認為是牛奶或雲呢嗱口味，其他口味尤其是水果口味的，我都覺得會搶去黑松露的獨特香氣。但一定要蛋糕取出後便要馬上放上雪糕，讓蛋糕的溫度使雪糕慢慢溶化，這樣才有「冰火熔岩」驚喜效果。

5. 蛋糕漿可以預先做好先放入容器再放在雪櫃保存，到上甜品前才放入焗爐，這樣在處理晚餐時便更加輕鬆，有更多的時間陪伴家人及賓客了。

沖繩黑糖年糕

Okinawa Kurozatou
New Year Cake...222

農曆新年是中國人的大節日，大家通常會到家人朋友家相聚，有人喜歡打牌聯誼，或者簡單聊天、吃吃賀年糕點的，熱熱鬧鬧，非常歡欣快樂。

每逢新年，都會想起小時候在外婆家吃她的自家製煎堆油角，而阿姨的新年拿手好菜則是年糕和馬蹄糕。今次是將小時候記憶中的美味年糕，變奏成為近年大受歡迎的「沖繩黑糖年糕」，更能為年糕多添一份黑糖的獨特香氣。

食得開心！

材料

- 沖繩黑糖............500 克
- 水.................1 公升
- 糯米粉.............830 克
- 油.................15 克

做法

1. 黑糖先用水煮溶，用密篩過濾雜質，涼卻後備用。

2. 糯米粉用粉篩過濾，將糯米粉逐些加入黑糖水，邊加邊攪拌成幼滑粉漿，加入油，繼續攪拌至油分與粉漿完全融合為止，再用密篩過濾雜質。

3. 在蒸盆內塗一層油，將粉漿倒入蒸盆，蓋上保鮮紙，水滾後放入蒸鑊蒸 90-100 分鐘，涼卻後放入雪櫃冷凍約兩天，讓年糕質地變硬定型後，便可切片，以慢火兩面煎香及軟身後便可享用了。

貼士

1. 黑糖水與糯米粉混合時一定要將糯米粉攪拌至完全融合，否則粉漿內尚有剩餘粉粒的話，蒸出來的年糕便不均勻及不夠細滑了。

2. 在粉漿內加入食油，目的是可令年糕的口感滑身一些，及質地光亮一些。

3. 每人所用的蒸盆厚度及家裏的火力可能都會不一樣，因此蒸糕的時間亦會可能適當加長或縮短一些。

4. 要知道蒸出來的年糕是否已經熟透，可用一枝筷子在年糕中央插入，如拔出來的筷子頭尚有較淺色的粉漿黏着的話，則代表年糕還未熟透，那蒸的時間便需要長一點了。

5. 蒸年糕前在蒸盆內塗抹一層油，可避免年糕黏着蒸盆，亦較容易將年糕移出。

私房南華李果酒

Homemade Chinese plum wine...223

這是一道很富特色，自家釀製的南華李果酒，蘊含南華李的獨有風味，很香、酸酸甜甜，雪凍後用作餐後甜酒，簡直無得彈！

私房南華李果酒的做法非常簡單，原理就像釀製葡萄酒，毋須添加任何酒精或酒的成分，只需將南華李剕開，混合冰糖，等待南華李果汁溶解冰糖，然後讓它靜止自然發酵而醞釀而成，一年後就自自然然變成有酒精的南華李果酒，過程就如經歷一趟葡萄園的醞釀之旅，非常神奇。

在香港，南華李的當造季節大約是七月，趁街市有得賣，大家都試試啦！

材料

- 南華李.....................10 磅
- 冰糖.....................8 磅
- 闊口密封玻璃瓶.....1 個

做法

1. 南華李洗淨，徹底瀝乾水分。

2. 將南華李水分抹乾，以小刀在南華李上剓開幾刀。

3. 將冰糖舂碎，然後一層李，一層冰糖裝滿容器。

4. 兩日之後，李子滲出的汁液混合冰糖，然後漸漸溶解。接着便等待它開始發酵。

5. 一星期後，果汁已經將冰糖完全溶解，液體已由最初的透明狀態轉化為較深的血紅色，瓶內正不斷有氣泡開始上升，這意味着液體已開始發酵，而且酵素相當活躍，這過程可能會持續幾個月，糖分在這個過程擔當一個很重要的角色，它是這瓶南華李果酒的催化劑，慢慢，經過發酵的液體就會轉化成酒精。

6. 整個醞釀過程大概歷時十個月至一年左右，這原本只是混合冰糖的南華李果汁，就會變成有酒精，充滿獨特南華李風味的南華李果酒。

貼士

在製作前一定要將南華李的水分抹乾，同時容器亦要經過高溫消毒及抹乾水分，不能有任何生水，否則很易會滋生細菌，令其發霉。

Zero-failure black garlic in rice cooker...10

INGREDIENTS
whole garlic cloves or single-clove garlic (with skin on)

UTENSILS
rice cooker
1 thick bamboo mat (about 0.5 cm thick, those made for teapots from homeware stores are preferred)
thin bamboo mats

METHOD
1. Put the thick bamboo mat into a rice cooker. Put a layer of garlic cloves over it. Top with a thinner bamboo mat. Arrange another layer of garlic cloves on top. Repeat this step by laying alternate layers of bamboo mats and garlic cloves until the rice cooker is full. Cover the lid and press "keep warm" button. Let the garlic cloves ferment in the rice cooker at about 60°C for 14 to 20 days. Do not open the lid throughout the process.
2. The garlic cloves will turn black and soft after fermentation. (If the garlic cloves are still chestnut brown after peeled, the fermentation hasn't completed.) The pungent spiciness of raw garlic will also be turned into sweetness. Remove the garlic cloves from the rice cooker. Leave them in a cool dry place for one or two days to air-dry. Store in airtight containers or zipper bags.

TIPS
1. For this recipe, do not rinse the garlic cloves before use. Do not open the lid throughout the fermentation process. Otherwise, the garlic may not ferment properly. Leave the rice cooker in an airy spot when the garlic is fermenting.
2. Pick garlic cloves that are plump without any sprout. I prefer garlic with purple skin because of its strong flavour. Single-clove garlic also works well because it has bigger cloves with a sweeter taste. Single-clove garlic can fill up the narrow spaces between layers of regular garlic cloves nicely.
3. For this recipe, it's advisable to use a rice cooker with a locking lid. That would prevent the lid from being opened accidentally.
4. I put a thicker bamboo mat on the bottom of the rice cooker. As the fermentation takes 2 to 3 weeks, the thick bamboo mat prevents the garlic from direct contact with the hot surface and getting burnt. Moreover, each layer of garlic cloves should be of equal thickness and try not to overlap them. That would ensure even temperature throughout the pot and let every garlic clove ferment the perfect way.
5. Humidity and temperature vary from season to season. Sometimes you may find a batch of black garlic a bit too moist after fermentation. When you make the next batch, you may put the fresh garlic cloves in a dehydration machine for a couple hours or dry them under the sun briefly before putting them into the rice cooker. The black garlic will turn out drier that way.

174

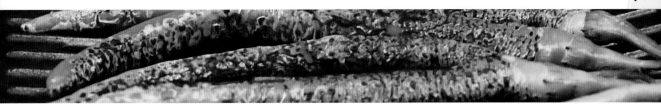

Homemade
black bean green chilli sauce...17

INGREDIENTS

1 kg medium-sized green chillies (I used Hunan chillies)
150 g shallot
50 g fermented black beans
300 ml vegetable oil
50 ml Sichuan peppercorn oil

SEASONING

4 tbsp sugar
3 tbsp white vinegar

METHOD

1. Rinse and wipe dry the green chillies. Preheat an oven to 250°C. Roast the chillies until slightly blackened and wrinkly. Let cool and carefully peel them. Finely chop the chillies and set aside. Rinse the shallot. Peel it and finely chop. Rinse the fermented black beans. Wipe dry and finely chop.

2. Heat vegetable oil in a wok. Fry the shallot until lightly browned. Add fermented black beans and stir well. Put in green chillies and toss again. Add Sichuan pepper oil and seasoning at last. Mix well and turn off the heat. Pour the chilli sauce into sterilized jars while still hot. Cover the lids and turn the jars upside-down to create a vacuum seal.

TIPS

1. The oven must be preheated first, so that the green chillies are exposed to extreme high heat. It's important to char and sear their skin within a short period of time. If they are left in the oven for too long, the chillies will be overcooked, soggy and discolour. The chilli sauce won't taste and look as good.

2. If you're not a big fan of fermented black beans, try satay sauce or fish sauce in place of them. The chilli sauce will render a different palate that is equally great. I've also tried putting in some chopped dried shrimps and shredded dried scallops. Feel free to give it a try and it's worth your while.

3. When you pick green chillies, look for those with thicker and plumper flesh. After the roasting step, much of the moisture is driven off. If you use chillies with little flesh, they tend to melt in the sauce and you won't be able to taste them. It's also advisable to remove some of the chilli seeds so that the sauce won't end up too gritty. I just don't like my chilli sauce to have too many seeds.

Homemade Renren sauce...20

INGREDIENTS

2.7 kg fresh Renren
200 ml white vinegar (for marinating Renren)
1.8 kg young ginger
60 g coarse salt (for marinating young ginger)
1.5 kg pork belly (with skin on)
320 g dried shrimps
620 g raw cane sugar slabs (diced) or
brown sugar
10 cloves garlic (sliced)
50 g bird's eye chillies (about 20 chillies, cut
into rings, the amount depending on your
tolerance to spicy food)
200 g white vinegar
850 g Liu Ma Kee fermented soy bean sauce
2 tsp salt (for seasoning)
1/4 cup Shaoxing wine
1 cup soaking water from dried shrimps
300 ml oil

METHOD

1. Rinse the Renren well. Cut off the stems.
   Make crisscross cuts on each. Remove the
   seeds. Add 200 ml of white vinegar and mix
   well. Leave them for 1 hour. Drain and set
   aside.
2. Rinse the young ginger and finely dice it.
   Add 60 g coarse salt and mix well. Leave
   them for 1 hour. Let the salt draw moisture
   out of the young ginger. Then drain and
   squeeze them well.
3. Rinse the pork belly. Burn off any hair on
   the skin with a kitchen torch. Then steam it
   with the skin on for about 50 minutes until
   done. Let cool. Dice finely with skin on.
4. Rinse the dried shrimps. Soak them in water
   for 1 hour until soft. Drain and set aside 1
   cup of the soaking water for later use. Finely
   chop the dried shrimps.

5. Heat a dry wok. Stir fry the diced young
   ginger until dry. Set aside.
6. Wipe dry the wok. Put in Renren. Fry them
   in the dry wok until yellowish brown. Set
   aside.
7. Wipe the wok clean. Add 300 ml of oil. Stir
   fry sliced garlic and chillies until fragrant.
   Put in dried shrimps and stir until fragrant
   and bubbly. Set aside.
8. Heat a bigger pot. Put in the pork and stir
   until oil starts to render. Add Shaoxing
   wine and stir further. Put in fermented
   soy bean sauce and raw cane sugar. Toss
   until the sugar melts. Put in Renren, young
   ginger, dried shrimps, bird's eye chillies,
   sliced garlic, 2 tsp of salt and the soaking
   water from dried shrimps. Keep stirring
   continuously for 10 minutes until the
   mixture thickens and turns glistening. Turn
   off the heat and let cool. Transfer into an
   airtight container. Store in the fridge.

TIPS

1. Renren sauce made with raw cane sugar
   slab tends to look darker in colour with a
   caramel-like aroma. Renren sauce made
   with brown sugar looks lighter in colour
   and tends to be thicker in consistency.
2. As the Renren sauce is thick and high in
   sugar content, you should keep stirring
   it while cooking. Otherwise, it might get
   burnt.
3. When the Renren sauce turns shiny, it
   is done and you may turn off the heat
   immediately. But if you want the Renren
   sauce thicker, you may keep on stir-frying it
   till the consistency you prefer.

Soy-poached oysters marinated in dill and olive oil...32

INGREDIENTS

24 oysters in shells (about 400 g)
dried small chillies
4 to 5 bay leaves
3 to 4 cloves garlic
30 g fresh dill
250 ml olive oil (enough to cover oysters)

SEASONING

3 tbsp sake (Japanese rice wine)
2 tbsp Mirin (Japanese cooking wine)
2 tbsp Maggi's seasoning
1 1/2 tbsp sugar

METHOD

1. Wipe down the dried chillies and bay leaves. Slice the garlic. Coarsely chop the dill. Mix the seasoning well. Shuck the oysters and discard the shells.

2. Put oysters into a non-stick pan. Bring to the boil. Water will come out of the oysters. Gently stir them until the liquid reduces and the oysters turn stiffer. Add seasoning and stir well. Keep cooking until the sauce reduces. Turn off the heat and set oysters aside to let cool.

3. Put the cooled oysters into a clean glass jar. Put dried chillies, bay leaves, garlic and dill over them. Pour in olive oil at last. Refrigerate for 2 days. Serve.

TIPS

1. Oysters come in different sizes. Anyway, the olive oil should be enough to cover all oysters.

2. You must get live oysters in shells for this recipe. Those packed in a tub taste fishy when cooked this way.

Home-style sun-dried food bursting with natural flavours dried cauliflower...14

INGREDIENTS

6 kg fresh cauliflower
3 tbsp coarse salt

METHOD

1. Cut cauliflower into florets. Boil water in a pot and add coarse salt. Divide the cauliflower into a few batches. Blanch each batch for 10-plus seconds in water. Drain.
2. Arrange the cauliflower evenly on a flat strainer. Leave the cauliflower in the sun for 1 week until completely dry. Store in zipper bags or airtight containers. Then keep in the fridge for later use.

TIPS

1. Do not cut the cauliflower into florets too small. Take into account the shrinkage when water is drawn from them. If you cut them too small, they will become tiny bits, not substantial enough for a good chew.
2. Dried cauliflower can be turned into a dish easily. Just rinse it and soak it in water until it swells and becomes soft. You can then make soup with it alongside pork ribs or other meat. It also goes well with braised meat, curry, stir-fried shredded pork and even a pot of congee.

NOTE

This recipe was published in my other cookbook *Cook it yourself@home*. As it is used in the next recipe *Braised pork belly with dried cauliflower and porcini mushrooms*, I include the method here once more so that you don't have to refer to another book back and forth.

Seafood salad with black garlic...24

INGREDIENTS

2 cloves black garlic (peeled, cut into quarters, mash some with a fork and use it in the dressing)
150 g salad greens
2 bananas (peeled, cut into short lengths)
2 eggs (hard-boiled, shelled, with egg whites dices and egg yolks mashed with a fork)
10+ medium prawns (blanched till done, shelled)
2 to 3 Hokkaido giant scallops (marinated with salt and ground white pepper, seared over high heat till half-cooked, cut into quarters)
ham
10+ cherry tomatoes (cut into quarters)
1 baby cucumber (sliced)

Black Garlic Vinaigrette Dressing

80 to 100 ml olive oil
1/4 to 1/2 onion (diced)
150 ml balsamic vinegar
mashed black garlic
freshly squeezed juice of 1 lime
dark brown sugar (to taste, crushed)

METHOD

1. Stir-fry onion in olive oil until soft. Set aside. Let cool and add other dressing ingredients. Mix well.
2. Put all ingredients (except mashed egg yolks and ham) into a salad bowl. Pour in dressing and toss well. Serve on a serving plate. Put the ham on top. Sprinkle with the mashed egg yolks. Serve.

TIPS

1. Feel free to improvise with your favourite ingredients. Use other seafood, meat and fruits as you wish.
2. You may use more or less black garlic according to your own taste and the size of the garlic cloves.
3. For better presentation, you may put mashed egg yolks, black garlic, prawns and scallops on top of the dressed leafy greens. You get to see the variety of ingredients used and it looks sumptuous.
4. The salad should be served right after tossed. Otherwise, the greens will wilt and lose their crunch. The water drawn out of the greens also make the dressing watery and unappetizing.

Hokkaido giant scallop sashimi with sea urchin cream sauce, crab roe and crabmeat...28

INGREDIENTS

12 sashimi-grade Hokkaido giant scallops
1 tray sea urchin (as topping)
salmon roe

Sea Urchin Cream Sauce

180 g sea urchin (about 4 trays)
180 ml milk
50 g whipping cream
1 tbsp oyster sauce

Crab Roe and Crabmeat

180 g freshly prepared crab roe and crabmeat
finely diced ginger
chicken stock
salt
caltrop starch slurry (as thickening glaze)

SEASONING

sea salt

METHOD

1. Thaw the scallops. Wipe dry with paper towel. Keep it in the fridge for later use.
2. Put all sea urchin cream sauce ingredients into a blender. Blend until smooth. Pass the mixture through a fine mesh to remove any lumps. Cover with cling film and keep in the fridge.
3. Heat some oil in a wok. Stir fry ginger, crab roe and crabmeat until fragrant. Add chicken stock and salt. Stir in caltrop starch slurry for thin glaze.
4. In a deep serving dish, pour in the sea urchin cream sauce. Slice each scallop into quarters. Put them over the sea urchin cream sauce. Then arrange some sea urchin, crab roe and crabmeat glaze and salmon roe over them. Sprinkle with sea salt and garnish. Serve.

TIPS

1. For food safety reasons, make sure all ingredients are kept in the fridge before serving or being cooked. Raw seafood also tastes better when chilled. If you have room in your fridge, it's best that you chill the serving dishes and cutlery too.
2. Apart from the crab roe and crabmeat glaze, you may also serve scallop sashimi with black truffle paste or XO sauce for variations that taste equally great.

Steamed fish head in Renren sauce...44

INGREDIENTS
1 head of bighead carp (about 600 g)
250 g homemade Renren sauce (see p.176 for method)
4 to 5 slices ginger
coriander or shredded ginger
3 tbsp oil

SEASONING
1/2 tsp sugar
1 tsp light soy sauce
1 tsp Shaoxing wine
1/2 tsp chicken bouillon powder
1/2 tsp caltrop starch

METHOD
1. Rinse the fish head. Drain well. Mix the seasoning together and rub the mixture on the fish head. Put some sliced ginger on a steaming plate. Put the fish head over the ginger and leave it for 15 minutes. Then spread the Renren sauce over the fish head.
2. Boil water in a steamer. Put in the fish and cover the lid. Steam over high heat for 10 to 12 minutes (depending on its size). Sprinkle with coriander or spring onion. Heat some oil until smoking hot. Drizzle over the fish. Serve.

TIPS
Optionally, you may put deep-fried tofu puffs or Chen Chuan rice noodles beneath the fish head. Such ingredients tend to pick up the flavours of the fish and Renren sauce well.

Cold tofu appetizer dressed in sea urchin soy sauce...34

INGREDIENTS
1 cube cold tofu (Momen tofu, silken tofu or any of your choice)
sea urchin
freshly picked crabmeat
salmon roe

Sea urchin soy sauce
80 g sea urchin
50 g soy sauce for sashimi
wasabi (Japanese horseradish)

METHOD
1. To make the dressing, put sea urchin, soy sauce and wasabi into a blender. Blend until smooth.
2. Arrange the tofu on a serving dish. Top with sea urchin, crabmeat and salmon roe. Drizzle with sea urchin soy sauce from step 1. Garnish and serve.

TIPS
1. The sea urchin soy sauce should be used immediately after blended. If there's any leftover, keep it in the fridge.
2. You may add wasabi according to your tolerance to spiciness.
3. As food is heated up in the blending process, make sure you keep the soy sauce in the fridge before using it to make the dressing. Otherwise, the bold flavour of fresh sea urchin may be undermined.

Thai-style Hokkaido giant scallop salad...37

INGREDIENTS

5 sashimi-grade Hokkaido giant scallops
2 stems lemongrass (use only the white part)
2 to 3 cloves garlic
2 sprigs coriander
6 Thai tomatoes
pomelo pulp
1 to 2 bird's eye chillies (amount depends on your tolerance for spicy food)
1 sprig Thai basil
3 shallots

MARINADE

salt
ground white pepper

DRESSING

3 tbsp lime juice
2 1/2 tbsp fish sauce
2 tbsp Thai palm sugar
1 tbsp sugar

METHOD

1. Thaw the scallops. Wipe dry with paper towel. Sprinkle with salt and pepper. Mix well and set aside. Slice lemongrass thinly. Slice the garlic and deep-fry until crispy. Cut coriander into short lengths. Cut each Thai tomato into quarters. Cut bird's eye chillies into rings. Pick the Thai basil leaves and discard the stems. Finely slice the shallot. Mix together all dressing ingredients until sugar dissolves.
2. Sear the scallops over high heat until both sides golden and browned. Cut each into halves. Put all ingredients (except the deep-fried sliced garlic) into a big salad bowl. Drizzle with the dressing and toss quickly. Save on serving plate. Sprinkle with deep-fried sliced garlic at last. Serve.

TIPS

1. Make sure you sear the scallops over high heat. Only strong heat can brown the scallops quickly while their centres are still raw, crispy and caramelized on the outside, juicy and tender on the inside. This is the best way to savour the delicate palate of Hokkaido giant scallops.
2. Apart from scallops, you may also use other seafood (such as squid or prawns) or meat of your choice. They taste equally great with the dressing.
3. The deep-fried sliced garlic should be sprinkled on top of the salad after it is dressed. Otherwise, the sliced garlic may pick up too much moisture from the dressing and won't be crispy.

Drunken hairy crabs with Shaoxing wine and dried plums...41

INGREDIENTS

20 hairy crabs (about 225 g each)
about 10+ perilla leaves

WINE MARINADE

1 litre Sorghum wine
2.4 litre Shaoxing wine
150 g dried plums
900 g sugar
1.9 litres light soy sauce
2 tbsp dark soy sauce
10 pods star anise
20 g cassia bark
10 bay leaves
10 g Sichuan peppercorns
10 g nutmeg
10 g dried bird's eye chillies
60 g spring onion
8 cloves garlic (gently crushed)
1 lemon (sliced)

METHOD

1. Rinse the hairy crabs. Put perilla leaves into the water in a steamer. Bring to the boil and put in the crabs. Steam over high heat for 15 minutes until cooked through. Set aside to let cool.
2. Rinse all the herbs and spices in the wine marinade. Drain and put all wine marinade ingredients into a pot. Bring to the boil and turn off the heat. Let cool.
3. Remove the straw strings on the crabs. Put them into the cooled wine marinade. Keep in a fridge for 24 hours. Serve.

TIPS

1. After marinated for 24 hours, the crabs will be flavourful enough. If you don't intend to serve at this point, please drain off the marinade. Otherwise, the crabs may get too salty if soaked in the marinade any longer. Over-seasoning the crabs would cover up their authentic flavour.
2. When you make the wine marinade, make sure you turn off the heat right after it comes to the boil. Sorghum wine and Shaoxing wine shouldn't be boiled for too long. Otherwise, their alcohol content will evaporate and their flavours will also be gone. Don't worry about whether the aromas of spices have enough time to infuse. Their flavours will keep on infusing as the marinade cools down and when the crabs are soaked.
3. You can keep the wine-marinated crabs in the fridge for 2 to 3 days. Yet, it's still advisable to consume as quickly as you can for food safety's sake.
4. When hairy crabs are not in season, you may use other crabs, such as swimmer crabs for this recipe. They taste equally great.

Steamed oysters with dried tangerine peel and olive black bean sauce...46

INGREDIENTS

12 U.S. oysters
finely chopped spring onion

OLIVE BLACK BEAN SAUCE

1 piece dried tangerine peel
50 g fermented black beans
50 g Chinese pickled olives
10 cloves garlic
2 to 3 bird's eye chillies (depending on your tolerance to spicy food)
150 ml oil

METHOD

1. To make the olive black bean sauce, soak the dried tangerine peel in water until soft. Rinse and scrape off the pith (the side with lighter colour) with a paring knife. Finely shred it and set aside. Rinse the fermented black beans and Chinese pickled olives. Wipe dry and finely dice them. Set aside. Finely chop the garlic and cut the bird's eye chillies into rings. Set side. Heat oil in a wok and stir fry garlic until golden. Put in all remaining sauce ingredients. Toss until fragrant. Turn off the heat. This is the olive black bean sauce.

2. Boil water in a steamer. Put the rinsed oysters in a microwave oven and heat it for 30 seconds on high power. Remove from microwave oven and put olive black bean sauce over them. Transfer into the steamer. Cover the lid and steam for 1 or 1 1/2 minutes (depending on the doneness you prefer and the sizes of the oysters). Sprinkle with finely chopped spring onion on top. Serve.

TIPS

1. Before you cook the oysters, make sure you keep them in the fridge for food safety reasons. Rinse them in ice water after taking them out and double check if there's any broken shell on them. Nothing annoys more than biting into broken shells when having your oysters.

2. Before steaming the oysters, par-cook them in a microwave oven for 30 seconds first. This steps helps seal in the moisture and juices of the oysters while bringing them to room temperature. When you steam them, they'll be cooked on the outside but still half-cooked and juicy on the inside. If you move the oysters straight from the fridge to the steamer, they'd take longer to cook. The prolonged cooking tends to draw water out of them and they won't stay plump.

3. Before you steam the oysters, make sure the water comes to a vigorous boil in the steamer. Start timing once you put the oysters in. If the oysters are in the steamer before the water boils, they have to be heated for a long time. It's hard to time and the oysters may be overcooked, without the tenderness of half-cooked ones.

Steamed swimmer crabs on egg white and coconut custard...49

INGREDIENTS

2 swimmer crabs (about 1.5 kg)
400 g coconut water
young coconut flesh
280 g egg white (about 8 egg whites)
finely chopped spring onion

SEASONING

5 g salt
5 g sugar

METHOD

1. Dress the crabs. Rinse well and chop into pieces. Drain. Bring water to the boil in a steamer. Steam the crabs over high heat for 6 minutes. Let cool. Drain any juices that came out of the crabs and set aside for later use.

2. In a mixing bowl, put in coconut water, egg whites, crab juices and seasoning. Whisk well. Pour into a deep steaming dish. Arrange the par-cooked crabs over. Cover with cling film. Bring water to the boil in a steamer. Steam the egg white custard and crabs over high heat for 8 minutes. Put in the young coconut flesh and cover with cling film again. Steam over high heat for 2 more minutes until the egg white has set and the crabs are cooked through. Sprinkle with finely chopped spring onion. Serve.

TIPS

1. As the crabs take longer to cook than the egg white, they need to be partly cooked before steamed with the egg white. That way, both ingredients will be cooked at the same time without overcooking the egg white custard, or undercooking the crabs.

2. Fresh young coconuts are available from shops specializing in spices and coconuts. Yet, they turn stale very quickly, especially in summer time. Make sure you keep them in the fridge. If you can't get fresh young coconuts, you may use bottled coconut water instead for similar results. But of course, you'd miss the tender and gelatinous young coconut flesh.

3. Young coconut flesh is sweet and creamy. It shouldn't be overcooked, or its tender texture will be lost. That's why I add them to the steamed custard two minutes before it's done.

Egg white omelette mille feuille topped with sea urchin and salmon roe...52

INGREDIENTS

300 g egg white (about 8 egg whites)
250 g milk
30 g sea urchin sashimi
salmon roe

SEASONING

25 g caltrop starch
2 tbsp water
4 g salt
3 g sugar
2 g chicken bouillon powder

METHOD

1. Mix caltrop starch with water to make a smooth slurry. Put in the rest of the seasoning ingredients and mix well.
2. Put egg whites and milk into a mixing bowl. Put in seasoning from step 1 and whisk well.
3. Heat some oil in a pan over medium-low heat. Swirl the pan to coat evenly. Pour out excess oil.
4. Pour in some egg white mixture from step 2. Swirl to coat evenly to make a thin omelette. Cook over low heat until it starts to set. Pleat the omelette by pushing it from right to left with a spatula. Transfer the omelette on a serving plate. Repeat the above step with the remaining egg white mixture and stack the omelettes on the plate. Arrange sea urchin, salmon roe and garnish over them. Serve.

TIPS

1. Adding caltrop starch to egg whites helps bind the omelettes better and they are not likely to turn wet after cooked. Add oil to the pan before frying each omelette because the oil not only prevents it from sticking to the pan, but also give it a lovely sheen and velvety texture. I mix the caltrop starch with water first before adding seasoning and egg whites. The batter will blend better that way.
2. Caltrop starch tends to sink quickly in the batter. Thus, make sure you whisk the mixture before pour it into the pan. Otherwise, your omelettes may not have even consistency.
3. To pleat an omelette nicely, you must push it with a spatula when it's half set. When the egg whites are cooked through and thoroughly set, it won't hold the pleats anymore.
4. When you fry the omelettes, make sure you use medium-low heat all the way. You can control the doneness and consistency of the egg whites more easily that way. If you fry them over high heat, they'll be tougher in texture. As the egg whites will be cooked in no time, it's also hard to grasp that exact moment to pleat them when they're still half-set.
5. Apart from sea urchin and salmon roe, I've also tried garnishing it with freshly picked crabmeat, black truffles and all kinds of mushrooms. They taste equally great.

Pepper salt-baked crab with Thai mint sauce...58

INGREDIENTS

2 male mud crabs (about 1.5 kg)
60 g white peppercorns
550 g coarse salt

Mint sauce

40 g fresh mint leaves
90 g freshly squeezed lime juice
45 g fish sauce
80 g sugar

METHOD

1. Pound the white peppercorns with mortar and pestle. Mix with coarse salt. Fry the mixture in a dry wok over medium-low heat until lightly browned and fragrant. Set aside.
2. Dress the crabs and keep them in whole. Separate the carapaces from the bodies. Line a baking tray with baking paper and put the crabs on. Cover with two sheets of mulberry paper (or baking paper). Wrap the crabs well. Then cover them evenly with the toasted pepper salt from step 1.
3. Preheat an oven to 250°C. Bake the crabs for 25 minutes. In the meantime, put all mint sauce ingredients into a blender. Blend until fine.
4. Carefully remove the pepper salt and mulberry paper. Chop the crabs into pieces and save on a serving plate. Serve with the mint sauce on the side as a dip.

TIPS

1. Do not chop the crabs into pieces. There are two reasons for that. First, it's harder to transfer them onto a serving plate if chopped up beforehand. Second, when you take the crabs out of the pepper-salt, you may get salt on them. If they are chopped up, they may get too salty.
2. Apart from crabs, you may also salt-baked giant prawns, clams and fish with the same recipe. Yet, make sure you adjust the baking time according to the size of the ingredients and how quickly they cook.

Hand-shredded chicken in spring onion ginger dressing...61

INGREDIENTS

1 chicken (about 1.5 kg)
8 to 10 slices ginger
5 to 6 sprigs spring onion
2 whole pods star anise
3 stems lemongrass (use only the white part)
3 to 4 pandan leaves
2 tbsp salt
1 tray ice water (enough to cover chicken)

SPRING ONION GINGER DRESSING

20 sprigs spring onion (about 300 g)
10 shallots (about 80 g)
3 knobs fresh sand ginger (about 40 g)
250 ml oil
10 g sea salt
5 g ground sand ginger
chicken bouillon powder

METHOD

1. Boil enough water to cover the chicken in a large pot. Put in ginger, spring onion, star anise, lemongrass, pandan leaves and salt. Stir until salt dissolves. Put in the chicken and soak for less than 20 seconds. Lift it out of the water. Rinse in cold water for 3 minutes.

2. Bring the same pot of water to the boil again. Put in the chicken and heat until the water boils again. Cover the lid and turn off the heat. Leave the chicken to soak for 45 minutes. Transfer chicken into an ice water bath. Leave it there for 20 to 30 minutes. Drain and tear the flesh off the bones. Save the shredded chicken on a serving plate.

3. To make the dressing, finely chop spring onion, shallot and fresh sand ginger. Heat some oil and stir-fry shallot and sand ginger until fragrant. Put in spring onion and stir briefly. Add sea salt, ground sand ginger and chicken bouillon powder. Toss well. Drizzle the hand-shredded chicken with the dressing. Serve.

TIPS

1. The chicken is steeped in ice water bath immediately after it's cooked. The cold water stops the cooking process right away, so that the chicken won't be overcooked by the remaining heat. Besides, the thermal shock also tightens the chicken skin texture to give it a gelatinous and crunchy mouthfeel.

2. Every chicken weighs and measures different. Thus, the soaking time in hot water may have to be lengthened or shortened accordingly. To test for doneness, insert a bamboo skewer into the fleshiest part on the chicken thigh. If the juices run clear (instead of bloody), it's done.

3. The water that the chicken was poached in can be used to make rice, blanch leafy greens, or used as a stock base for making soups. It's too tasty to be poured down the drain.

Fried dace tofu strips in black truffle sauce...64

INGREDIENTS

3 cubes cloth-wrapped tofu
150 g minced dace
30 g large dried shrimps
1/2 egg
20 g caltrop starch

SEASONING

1/2 tsp salt
1/2 tsp sugar
1/2 tsp chicken bouillon powder
ground white pepper
sesame oil

Black truffle sauce

35 g frozen black truffle
olive oil
1/4 tsp salt

Caltrop starch slurry

(as batter for coating minced dace tofu strips before frying)
25 g caltrop starch
3 g chicken bouillon powder
30 g water

METHOD

1. Put tofu into a fine mesh strainer. Mash with your hands and drain well (too much moisture in tofu will stop the minced dace mixture from binding properly). Set aside. Soak dried shrimps in water till soft. Drain. Fry in a wok with a little oil until they curl up and smell fragrant. Drain the oil and finely chop the dried shrimps. Set aside.
2. In a mixing bowl, stir together mashed tofu, minced dace, dried shrimps, egg, caltrop starch and seasoning. Stir until sticky.

3. Grease a rectangular pan lightly. Put in the tofu mixture. Level the surface. Cover in cling film. Boil water in a steamer. Steam dace tofu mixture over high heat for 18 to 20 minutes (depending on its thickness). Let cool.
4. To make the black truffle sauce, finely chop the black truffle. Add olive oil and salt. Stir well and set aside for later use. Mix the caltrop starch slurry ingredients together. Set aside.
5. Refrigerate the steamed dace tofu patty for 12 hours until fully set.
6. Cut the steamed dace tofu patty into strips. Dip into the caltrop starch slurry. Shallow-fry over medium heat until both sides golden. Save on a serving plate. Drizzle with some black truffle sauce. Garnish and serve.

TIPS

1. The dried shrimps added to the dace tofu mixture lend this dish an extra aroma and seafood flavour. Apart from stir-frying them in a little oil, you may also roast them in an oven or deep-frying them before use.
2. I coated the dace tofu strips in caltrop starch slurry before frying them. That way, the dace tofu strips will be crispier.
3. I used cloth-wrapped tofu from wet market for this recipe because of its denser and firmer texture. As the dace tofu strips have to be fried at last, I want them to hold their shapes better. But if you're making to traditional steamed dace tofu, you may use pre-packed soft tofu from supermarkets instead for silkier and softer texture.

Avocado au gratin with seafood and cheese...68

INGREDIENTS

3 ripe avocadoes
6 medium prawns
5 to 6 scallops
100 g squids
6 to 8 strawberries
1/2 granny smith apple
25 g grated parmesan cheese
120 g grated mozzarella cheese
100 g creamy salad dressing
salmon roe

MARINADE FOR SEAFOOD

salt
ground white pepper

METHOD

1. Rinse the avocadoes and wipe dry. Cut each in half. Remove the pit. Rinse all seafood and fruits. Wipe dry and dice separately.
2. Add salt and ground white pepper to the diced seafood. Mix well and leave it briefly. Heat oil in a wok. Stir-fry the seafood until medium-well done. Drain off any liquid.
3. Let the seafood cool for a while. Add diced strawberries, apple, parmesan, creamy salad dressing and 1/4 of the mozzarella cheese. Mix well and stuff the avocadoes with the mixture. Then sprinkle the remaining mozzarella evenly on them. Put the stuffed avocadoes on a baking tray.
4. Preheat an oven to 200°C. Bake the avocadoes for 10 to 12 minutes until the cheese on top has melted and lightly browned. Garnish with salmon roe. Serve.

TIPS

1. Do not cook the seafood for too long in step 2. Otherwise, it will be overcooked after baked and become tough and rubbery. Besides, make sure you drain any liquid from the seafood before mixing it with other filling ingredients. Otherwise, the filling may become too watery.
2. After cutting the avocadoes in half, you may cut off 2 mm from the base so that they can rest on the flat surface. Then the avocadoes won't topple on the baking tray and serving plate.
3. Adding parmesan cheese to the salad dressing gives the filling a fragrant cheesy flavour, whereas the mozzarella helps bind the filling and makes it stringy.
4. Feel free to use other seafood or fruits of your choice for variations.

Tom Yum roast chicken...71

INGREDIENTS
1 chicken (about 1.2 kg)

MARINADE
90 g Tom Yum paste (a)
40 g Tom Yum paste (b)
25 g grated garlic
25 g chopped shallot
20 g galangal (finely chopped)
2 sprigs coriander (with roots, finely chopped)
2 stems lemongrass (use only the white part, finely chopped)
8 to 10 Kaffir lime leaves (finely chopped)
50 ml water
3 tbsp oil
caltrop starch slurry (for thickening the sauce)

SEASONING
1/4 tsp salt
1/2 tsp sugar
1/2 tsp chicken bouillon powder

METHOD
1. Rinse the chicken and wipe dry with paper towel. Rub 90 g of Tom Yum paste on both the insides and the outsides of the chicken. Set aside.
2. Heat oil in a wok. Stir-fry garlic and shallot until fragrant and lightly browned. Add galangal, coriander, lemongrass and Kaffir lime leaves. Stir well. Put in 40 g of Tom Yum paste. Stir again. Add water and seasoning. Mix well. Thicken the sauce with caltrop starch slurry. Set aside to let cool completely. Stuff the chicken with this aromatic and herb mixture. Seal the opening with metal skewer. Wrap in cling film and leave it in the fridge overnight.
3. Preheat an oven to 150°C. Roast the chicken for 60 minutes. Turn the oven up to 250°C and roast for 15 to 20 more minutes until the skin is crispy and golden.
4. Tear the meat off the bones with your hands. Mix the chicken meat with the aromatic herb mixture. Serve.

TIPS
1. To make roast chicken with crispy skin and succulent meat, make sure you pick a chicken with enough fat. Throughout the 30-minute roasting process, the subcutaneous fat has enough time to melt and crisp up the skin. Lean chicken tends to be dry and chewy after roasted.
2. Before marinating the chicken, wipe it completely dry on both the inside and the outside. Any residual moisture on the skin would dilute the marinade and make the chicken less flavourful.
3. Tom Yum paste is the key component in Thai Tom Yum Goong soup and is available from grocery stores specializing in Thai ingredients. There are different brands and they taste quite different in terms of sourness and spiciness. Please feel free to sample different brands to look for your preferred one.
4. When you make the aromatic mixture to be stuffed in the chicken, you may make the sauce thicker than usual. It's easier to handle that way.

Atlantic green whelks in spicy wine sauce with Thai basil...74

INGREDIENTS

1.8 kg frozen Atlantic green whelks
60 g grated garlic
60 g finely chopped shallot
100 g diced ginger
20 g fresh bird's eye chillies (finely chopped)
1 litre chicken stock
20 g ground dried bird's eye chilli
10+ bay leaves
6 slices liquorice
Kuzukiri (Japanese kudzu noodles)
5 to 6 sprigs Thai basil (leaves only)

SEASONING

500 ml Shaoxing wine
200 ml Chinese rose wine
80 g chilli oil
60 g Guilin chilli sauce
70 g sesame oil
50 g satay sauce
80 g fish sauce
2 tsp sugar

METHOD

1. Thaw the whelks and rinse well. Put them into a pot. Add enough cold water to cover. Bring to the boil and cook for 30 minutes. Turn off the heat and cover the lid. Wait till the whole pot of water has cooled completely. Drain. Remove the operculum on each whelk.

2. Heat a pot and add 4 to 5 tbsp of oil. Stir fry grated garlic, shallot and ginger. Put in fresh bird's eye chillies and stir fry until fragrant. Add chicken stock, dried bird's eye chilli powder, bay leaves and liquorice. Bring to the boil over high heat. Turn to medium-low heat and cook for 15 minutes (without covering the lid) until the flavours are infused.

3. Put in all seasoning. Bring to the boil again. Cook over medium-low heat for 5 to 10 more minutes so that the alcohol is evaporated. You may cook it longer or shorter depending on your personal preference on the alcohol taste.

4. Put in the whelks. Bring to the boil again and turn off the heat. Let cool and refrigerate to let the flavours penetrate.

5. Before serving the next day, soak the Kuzukiri in water till soft. Boil in water until it turns transparent. Drain and put it on the bottom of a deep serving dish. Reheat the whelks in the spicy wine sauce. Put in Thai basil leaves at last and stir well. Cook till the basil leaves wilt. Pour the mixture into the serving dish over the bed of Kuzukiri. Serve.

TIPS

1. This dish is spicy and very appetizing. It's the perfect companion to alcoholic drinks. You may adjust the amount of hot spices and chillies used according to your tolerance to spicy food.

2. Apart from whelks, you may also use prawns, crabs or clams instead. Just make sure you serve them right away after the seafood is cooked in step 4. You don't have to let them sit overnight to pick up the flavours.

3. The sauce itself is very spicy and delicious. You can actually turn it into a hotpot soup base. You can use the leftover sauce to cook instant noodles, mung bean vermicelli or ribbon rice noodles. Any staples that pick up the sauce will work fine.

Top-secret tomato beef curry...78

INGREDIENTS A
1.5 kg beef shin
2 onions
3 to 4 potatoes (peeled and cut into wedges)
500 g canned peeled tomatoes (coarsely chopped)
3 fresh tomatoes (cut into pieces)

INGREDIENTS B
450 g yellow curry paste
1.5 litre beef stock
3 stems lemongrass (bruised with the back of a knife)

SEASONING
80 g fish sauce
50 g sugar
200 g ketchup
1/2 cup coconut milk

METHOD
1. Blanch the beef shin in boiling water. Drain and transfer into a soup pot. Add 2 litres of water and boil for 1 hour. Drain and set aside the beef stock for later use. Let cool and slice beef into chunks.
2. Dice the onion. Fry the potato in oil until lightly browned.
3. Heat 3 tbsp of oil in a wok. Stir-fry onion until soft. Add curry paste and stir until fragrant. Add beef stock from step 1, canned tomatoes, lemongrass, fish sauce, sugar, ketchup and beef shin. Cook over low heat for 30 minutes. Add potatoes and keep on cooking for 15 minutes. Put in the fresh tomato at last. Cook over low heat for 15 more minutes. Turn off the heat and cover the lid. Let the curry cool down. This curry tastes even better the next day. Before serving, add coconut milk and bring to the boil. Or heat up the curry and stir in coconut milk at last.

TIPS
1. Whenever you use beef shin in dishes with dense and thick sauce, it's advisable to par-cook the beef shin in water for 1 hour first. This step shortens its cooking time with other seasonings and ingredients. Otherwise, all remaining ingredients will be mushy and the sauce will be too thick and rich after hours of cooking. It's harder to control the taste and consistency of the end product.
2. To accentuate tomato taste in a dish, add canned tomatoes and ketchup on top of fresh tomatoes.
3. You can get yellow curry paste from supermarkets and shops specializing in coconut milk and spices.

Marinated pork, eggs and tofu in "Dai Pai Dong" style...82

INGREDIENTS

1.2 kg pork belly
8 to 10 cloves garlic (crushed)
5 pieces deep-fried tofu
8 hard-boiled eggs (shelled)

SPICED MARINADE

4 pods star anise
1 small section cassia bark
1 tbsp white peppercorn
8 coriander roots (crushed)
3 to 5 bird's eye chillies or dried red chillies
(use more or fewer according to your tolerance
for spicy food)
6 tbsp dark thick soy sauce
8 tbsp fish sauce
60 g palm sugar
8 cups water

METHOD

1. Rinse the pork and cut into 1 1/2-inch cubes. Wipe dry.
2. Heat 3 tbsp of oil in a wok. Stir fry garlic over low heat until golden. Add pork and turn to high heat. Stir fry until the pork turns white and lightly browned.
3. To make the marinade, put all spices into a small muslin bag. Tie a knot to secure. Put it into a pot. Add the remaining marinade ingredients. Put in pork, deep-fried tofu and hard-boiled eggs. Bring to the boil over high heat. Turn to low heat and simmer for 1 1/2 hours (without covering the lid) until the pork belly is tender.

4. For serving at once, turn the heat to high and cook until the marinade thickens and serve. Alternatively, you may turn off the heat after the pork is cooked through. Leave all ingredients in the pot with the lid covered overnight. Reheat before serving the next day. It would taste even more flavourful.

TIPS

1. Dark thick soy sauce is a common seasoning in Southeast Asia. It's dark in colour with a characteristic caramel-like sweetness. It's thicker and more flavourful than regular soy sauce. Commonly served as a dip on the side with food like Hainan Chicken Rice or soup ingredients, dark thick soy sauce can also be used in stir-fried, boiled, braised, marinated dishes and cold appetizers. You can get it from shops specializing in spices and coconut, Thai grocery stores and even some large-scale supermarkets.
2. When you cook the pork, eggs and tofu, the lid should never be covered. That would boil off the water so that the marinade gets thicker with time. The marinated food and the marinade itself will be more flavourful that way.
3. If you can afford the time, toast the spices (e.g. star anises and cassia bark) in a dry wok over low heat until fragrant before using. That would draw out their aromas and give the marinade a more mellow complex palate.

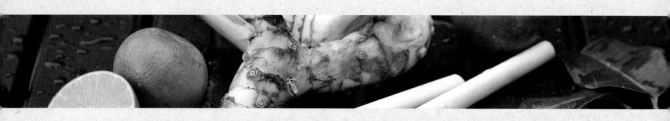

Thai-style roast chicken...84

INGREDIENTS

1 chicken (about 1.5 kg)

MARINADE

2 1/2 tbsp oyster sauce
2 tsp sesame oil
2 tsp Maggi's seasoning
1/2 tsp five-spice powder
2 tsp palm sugar
1 tsp chicken bouillon powder
1 1/2 tsp black peppercorns (coarsely crushed)
5 to 8 cloves Thai garlic (crushed with skin on)
3 coriander roots (crushed)

Khao Khua (Toasted sticky rice powder)

60 g glutinous rice
3 to 4 Kaffir lime leaves (with the centre veins torn off)
1 stem lemongrass
10 g galangal

Roast chicken sauce

2 tsp finely chopped Thai chillies
2 tsp palm sugar
1 1/2 tbsp fish sauce
1 tbsp lime juice
2 tsp tamarind juice
1 tsp Khao Khua
coriander (finely chopped)
spring onion (finely chopped)
mint leaves (finely chopped)
shallot (finely chopped)

METHOD

1. To make Khao Khua, finely chop the Kaffir lime leaves, lemongrass and galangal. Mix in the glutinous rice. Fry the mixture in a dry wok over low heat until lightly browned and fragrant. Let cool and pound into fine powder with mortar and pestle.

2. Rinse the chicken and chop into halves. Wipe dry with paper towel. Mix the marinade well and rub it on both the insides and outsides of the chicken evenly. Wrap in cling film and leave it in the fridge overnight.

3. Mix the roast chicken sauce ingredients. Preheat an oven to 185°C and roast the chicken with the skin side up for 40 to 45 minutes until golden and crispy. Chop or tear into pieces. Serve with the roast chicken sauce on the side as a dip.

TIPS

1. Before roasting the chicken, try to pull the thighs gently away from the breast. That would relax the meat and make it cook more easily.

2. Apart from roasting in an oven, you may also grill the chicken over charcoal or chop it up and coat it in caltrop starch for deep frying. It tastes equally great. If you feel like taking the easy way out, use whole chicken wings instead.

3. Khao Khua is an essential condiment in North-eastern Thai cuisine that gives food a smoky fragrance and crunchy texture. It is commonly used in Thai sauces, lettuce wraps and salads.

Kimchi scallop roll au gratin...88

INGREDIENTS

6 sashimi-grade Hokkaido giant scallops
1 sprig whole-leaf cabbage Kimchi
cheddar cheese
salmon roe
bacon bits (toasted)
2 tsp salt

UTENSILS

cooking thermometer
kitchen torch

METHOD

1. Thaw the scallops. Fill a pot up to 2/3 its height. Add 2 tsp of salt and stir well. Put in the cooking thermometer. Heat the water up to 70°C over low heat. Put in the scallops. Soak them for 2 to 2 1/2 minutes. Drain and dunk them into ice water. Let chill for 2 minutes. Drain again. Wipe dry with paper towel.

2. Lay flat a piece of cabbage kimchi. Put the scallop on top. Fold the Kimchi to wrap the scallop. Transfer onto a serving plate. Chop or grate some cheddar cheese on top of the Kimchi roll. Melt the cheese with a kitchen torch over low heat. Garnish with salmon roe and sprinkle with some bacon bits. Serve.

TIPS

1. Sashimi-grade scallops are meant to be eaten raw for its juicy texture. Thus, do not soak the scallops in warm water for too long. They'd be rubbery if cooked through.

2. Before you wrap the scallops in Kimchi, make sure you wipe the scallops dry with paper towel. Otherwise, the moisture from the scallops will seep through the Kimchi after wrapped, ruining the presentation and flavours.

3. You can get sashimi-grade Hokkaido scallops from large-scale Japanese supermarkets or shops specializing in imported Japanese frozen food.

Chive oyster fritter...91

INGREDIENTS

700 g U.S. oysters in tub
300 g Chinese chives
150 g sweet potato starch
150 g flour
2 eggs
600 ml water

SEASONING

1-1/2 tsp salt
ground white pepper

METHOD

1. Rinse the oysters. Drain and cut into pieces. Rinse the Chinese chives. Drain and dice them.
2. In a mixing bowl, put in sweet potato starch, flour, eggs, Chinese chives and seasoning. Stir in water and mix into a smooth batter. Lastly add oysters and stir a couple times to distribute well.
3. Heat oil in a non-stick pan. Pour in a small ladle of batter. Fry until both sides golden. Transfer onto paper towel to pick up excess oil. Save on a serving plate.
4. Serve with chilli sauce, fish sauce or XO sauce if you like.

TIPS

1. Oysters that come in a tub is cleaner than raw ones in shells. Just rinse them with water briefly. However, if you get oysters that are shucked live from wet market, you have to clean them more meticulously. Put the oysters into a bowl. Add caltrop starch and rub them gently. The dirt will adhere to the caltrop starch. Then rinse the starch away.
2. Do not dice the oysters too finely. Otherwise, you don't get to taste their creamy texture.
3. You may want to use a bit more oil for frying than usual for this recipe. The oyster fritters will be crispier and fluffier if you fry them in half-shallow half-deep manner. Besides, do not fry them over low heat. Otherwise, the batter will pick up too much oil and the fritters will be too greasy.
4. Add oysters to the batter only after the batter has been properly stirred. The oysters will break down in bits and pieces if stirred for too long.
5. You can add more or less water to the batter according to your preference. If you want your oyster fritters to be stiffer and thicker, use less water. If you want them thinner and moister, use more.

Poached beef in Sichuan "Shuizhu" style...94

INGREDIENTS A

4 cloves garlic (crushed)
40 g ginger (sliced)
150 g Sanwu Ma La hotpot base
80 g Sichuan spicy bean sauce
20 g dried chillies
12 g Sichuan peppercorns
1 litre chicken stock
70 g butter

INGREDIENTS B

50 g Kuzukiri (Japanese kudzu noodles)
500 g fresh beef tenderloin
20 g caltrop starch
30 g water
300 g pork blood curd
100 g soybean sprouts
baby cucumber (sliced)

SEASONING

10 g Shaoxing wine
1 tsp chicken bouillon powder
3 tsp sugar

GARNISHES

toasted sesames
2 to 3 sprigs coriander (cut into short lengths)

METHOD

1. Soak Kuzukiri in water over night. Drain.
2. Slice the beef tenderloin. Mix caltrop starch with water and mix into a slurry. Put in the beef and mix well.
3. Heat a pot and add a little oil. Stir-fry garlic and ginger until fragrant. Add 10 g of dried chillies, Sichuan spicy bean sauce, Sanwu Ma La hotpot base and 6 g of Sichuan peppercorns. Stir until fragrant. Sizzle with Shaoxing wine. Add chicken stock, chicken bouillon powder and sugar. Bring to the boil.
4. Put in pork blood curd. Cook for 3 to 4 minutes. Add soybean sprouts, cucumber and Kuzukiri. Cook for 2 to 3 minutes. Transfer the solid ingredients into a big serving bowl with a strainer ladle.
5. Boil the soup base in the pot again. Turn off the heat and stir in the beef tenderloin coated in caltrop starch slurry. Leave it in the soup until medium-well done. Pour both the beef and the soup into the serving bowl with other solid ingredients.
6. Heat some oil in a wok. Fry the remaining 10 g of dried chillies and 6 g of Sichuan peppercorns until fragrant. Pour the hot mixture over the beef tenderloin. Sprinkle with sesames and garnish with coriander. Serve.

TIPS

1. Sanwu Ma La hotpot base is a pre-packed condensed soup base. It's perfect and convenient for "Shuizhu" style dishes and Sichuan hotpot. You can get it from grocery stores in wet markets.
2. Apart from beef, you can also make fish, frogs, meat and seafood the same way. They taste equally great.
3. You may adjust the amount of spicy ingredients used according to your tolerance to spicy food.

Roast chicken stuffed with Kimchi and Korean rice cake...98

INGREDIENTS
1 chicken (about 1.5 kg)
70 g Gochujang (Korean chilli paste)
80 g Korean rice cake
220 g cabbage Kimchi
8 g salt

METHOD
1. Rinse the chicken and wipe dry. Soak the Korean rice cake in water until soft. Drain.
2. Rub Gochujang on both the inside and outside of the chicken. Finely chop cabbage Kimchi. Mix Kimchi with salt and Korean rice cake. Stuff the chicken with them. Seal the seam with a metal skewer. Wrap in cling film and leave it in the fridge overnight.
3. Preheat an oven to 150°C. Bake the chicken for 60 minutes. Turn the oven up to 250°C and bake for 15 to 20 minutes until the skin is golden and crispy. Serve.

TIPS
1. To make roast chicken with crispy skin and succulent meat, make sure you pick a chicken with enough fat. Throughout the 30-minute roasting process, the subcutaneous fat has enough time to melt and crisp up the skin. Lean chicken tends to be dry and chewy after roasted.
2. Before marinating the chicken, wipe it completely dry on both the inside and the outside. Any residual moisture on the skin would dilute the marinade and make the chicken less flavourful.

Steamed grouper with black garlic, ham and dates...55

INGREDIENTS
1 grouper (about 600 g to 900 g)
10 black garlic (coarsely chopped)
1 small piece dried tangerine peel (soaked in water till soft, rinsed and shredded)
50 g Jinhua ham (diced)
40 g ginger (diced)
4 black dates (rinsed, pitted and diced)
3 sprigs spring onion (finely shredded)

SEASONING
soy sauce for steamed fish
oil

METHOD
Dress the grouper and rinse well. Wipe dry. Put on a steaming plate. Arrange all remaining ingredients (except spring onion) over the fish evenly. Boil water in a steamer. Steam the fish over high heat for 10 to 12 minutes. Sprinkle with spring onion all over. Drizzle with soy sauce for steamed fish. Heat oil until smoking hot. Pour over the spring onion and the fish. Serve.

TIPS
I've tried steaming giant grouper belly instead of whole grouper fish. Not only is the grouper belly fleshier without worries of choking on bones, it is also rich in collagen because of its thick gelatinous skin. The fish oil and black garlic also match each other well.

Steamed pork ribs in chilli black bean sauce~Secret tricks from professional chefs...101

INGREDIENTS
600 g fresh pork belly ribs

CHILLI BLACK BEAN SAUCE
(This amount is enough for 1.2 kg of pork ribs. Use half it and save the leftover for next time.)
80 g fermented black beans
1 to 2 bird's eye chillies
3 cloves garlic
2 shallots
1 piece dried tangerine peel
1 tsp sugar

SEASONING
1 tsp caltrop starch
1/3 tsp salt
3/4 tsp sugar
1/2 tsp chicken bouillon powder
2 tsp garlic oil

METHOD
1. Rinse the pork and wipe dry. Rinse the black beans and wipe dry. Cut chillies into rings. Finely chop garlic and shallot. Soak dried tangerine peel in water until soft. Rinse and scrape off the pith. Finely shred it.
2. Heat 3 tbsp of oil in a wok. Stir-fry chillies, garlic, shallot and dried tangerine peel until fragrant and lightly browned. Add fermented black beans and sugar. Stir-fry for one more minute. Set aside. In other wok, boil water and steam the mixture for 20 minutes. This is the chilli black bean sauce.
3. Take 2 tbsp of chilli black bean sauce and mash it with a spoon. Add seasoning and mix well. Pour the mixture in with the pork belly ribs. Mix well and leave them for 1 hour.
4. Boil water in a steamer or wok. Put in the ribs and steam over high heat for 15 minutes. Sprinkle with finely chopped spring onion on top. Serve.

TIPS
1. Fry the chilli black bean sauce ingredients first. Then steam the mixture. The black beans will loosen up their structure and release more flavour that way.
2. Garlic oil is the oil that has been used to fry garlic. It's bursting with garlic flavour. Not only does it add aroma to the ribs, but also keep the ribs succulent and tender.
3. Pork belly ribs are close to the cartilage with some fat attached. They have soft texture and are tender and juicy after steamed.

Homemade crispy roast pork belly...104

INGREDIENTS

1 piece pork belly (about 1.8 kg)

MARINADE

2 tbsp Chinese rose wine
1 tbsp Shaoxing wine
1 tsp ground white pepper
1 tsp five-spice powder
3 tbsp coarse salt A
2 tbsp white vinegar
350 g coarse salt B

UTENSILS

pork skin pricker
aluminium foil

METHOD

1. Rinse the pork belly. Wipe dry and de-bone it (if there's any bone). Mix Chinese rose wine and Shaoxing wine in a small bowl. Rub the wine mixture evenly on all sides of the pork. Then rub pepper, five-spice powder and coarse salt A over it. Put it into a container with lid with the skin facing up. Keep it in the fridge overnight.

2. Prick holes all over the skin with a pricker. Put it on a sheet of aluminium foil. Fold the four sides up so that it looks like a tray. Then brush white vinegar on it and cover it with coarse salt B.

3. Preheat an oven to 200°C. Bake the pork for 50 minutes. Remove the coarse salt on the pork and pork skin. Prick holes all over the skin with the pricker again.

4. Turn the oven up to 270°C. Put the pork directly on a wire rack. Bake for 30 minutes until the pork skin puffs up and blisters. Remove from oven to cool briefly. Cut into pieces and serve.

TIPS

1. Burn off any hair on the skin with a kitchen torch if there's any before use.

2. Do not serve the pork straight out of the oven. Let it sit for a while to cool off a little and for the moisture to evaporate. The skin will be crispier that way.

3. When you prick holes on the pork skin, make sure you prick every inch of it. Those holes make the skin puff up and introduce air which will become blisters when baked. This is the key to crispy pork skin.

4. Any leftover roast pork belly will not have crispy skin the next day. Don't be wasteful and steam the leftover with fermented shrimp paste the next day. Or stir-fry it with fermented shrimp paste, sugar and Chinese chives. That would give the leftover a new lease of life by turning it into a delicious dish.

Ma La salted chicken salad with cucumber, coriander and shrimp roe...108

INGREDIENTS

2 fresh chicken legs (with thighs)
1 tbsp Sichaun peppercorns
1 tbsp coarse salt
4 to 5 sprigs coriander
1 baby cucumber
dried shrimp roe
toasted sesames

SEASONING

1 1/2 tbsp sesame oil
3/4 tbsp Sichuan pepper oil

METHOD

1. Rinse the chicken legs and wipe dry. Set aside. Mix Sichuan peppercorns with coarse salt. Stir-fry in a dry wok over low heat until lightly browned and fragrant. Let cool.
2. Rub the Sichuan pepper salt from step 1 on the chicken legs. Wrap in cling film. Refrigerate overnight.
3. Rinse off the Sichuan pepper salt on the chicken legs. Set aside. Boil water in a steamer and steam the chicken legs over high heat for 15 minutes until done. Drain any liquid on the plate. Leave the chicken legs in a breezy spot for 2 hours to cool and dry them thoroughly.
4. Rinse the coriander and baby cucumber. Drain. Cut coriander into short lengths. Cut the baby cucumber into irregular wedges while rolling it on the chopping board. Tear the chicken meat off the bones. Put all ingredients into a large bowl. Drizzle with sesame oil and Sichuan pepper oil. Toss well and save on a serving dish. Sprinkle with shrimp roe and toasted sesames. Serve.

TIPS

1. Make sure you wipe the chicken legs dry before marinating it with Sichuan pepper salt. Otherwise, the moisture will dilute the marinade and the flavours won't be intense enough.
2. After the chicken legs are steamed, they should be allowed to cool and air-dry. This firms up the chicken flesh and dries any moisture on the surface. Otherwise, the salad will be too wet after tossed. Besides, the chicken flavour is intensified after dried, complementing nicely with the tangy refreshing palate of coriander and baby cucumber.
3. To heighten the aroma of shrimp roe and to make it last longer, you may stir-fry it in a dry wok over very low heat with sliced garlic and shredded dried tangerine peel that has been soaked in water. When the dried tangerine peel and sliced garlic dry up, pass the mixture through a sieve. Let the shrimp roe cool completely and store in an airtight jar.
4. Such chicken legs marinated with Sichuan pepper salt also make great Hong Kong style clay pot rice alongside Cantonese preserved pork sausage, liver sausage or preserved pork belly.

Bone-softened Pacific saury in tomato sauce...111

INGREDIENTS

6 Pacific sauries (or Sanma in Japanese)
5 shallots (finely chopped)
5 cloves garlic (finely chopped)
1 to 2 bird's eye chillies
3 to 4 sprigs spring onion (cut in half, then into short lengths)

Tomato sauce

500 g tomato puree
3 bay leaves
3 tsp fish sauce
6 tsp sugar
1 tsp salt
2 tsp chicken bouillon powder
2 tbsp Tabasco sauce
3 tbsp Worcestershire sauce
juice of 1 lime
ground black pepper

METHOD

1. Thaw and dress the sauries. Remove all innards and rinse well. Wipe dry. Cut each in half across the length. Set aside.
2. Heat some oil in a wok. Stir-fry shallot and garlic until fragrant. Arrange the spring onion in the wok. Add bird's eye chillies and tomato sauce ingredients. Put in the sauries at last. Cook over high heat until it boils. Cover the lid and turn to low heat. Simmer for 2 hours until the bones are softened. Turn off the heat and keep the lid on. Let the wok cool down. Store in a fridge to let flavours mingle. Reheat and serve the next day.

TIPS

You may also cook sardines with the same method. Feel free to use more or less bird's eye chillies according to your tolerance for spicy food.

Braised pork belly with dried cauliflower and porcini mushrooms...114

INGREDIENTS
1.2 kg pork belly
70 g dried porcini mushrooms
70 g dried cauliflower (see p.178 for method)
water

SEASONING
150 g oyster sauce
35 g Chu Hau sauce
35 g ground bean sauce
35 g fermented red taro curd
35 g fermented beancurd
35 g sesame paste
dark soy sauce
70 g rock sugar

METHOD
1. Rinse the pork belly. Burn off the hair with a kitchen torch. Blanch in boiling water in whole. Steam over high heat for 45 minutes. Soak in ice water until cool. Cut into 2-inch cubes.
2. Rinse the porcini mushrooms. Soak in water for 12 hours until soft. Drain and set aside the soaking water for later use.
3. Rinse the dried cauliflower. Blanch in boiling water briefly. Drain.
4. Heat oil in a wok. Stir-fry seasoning until fragrant. Put in the pork belly and fry until lightly browned. Add water to almost cover the pork. Bring to the boil over high heat. Cook for 10 minutes. Turn to low heat and simmer for 45 minutes. Put in the porcini mushrooms, dried cauliflower and the water used to soak the porcini. Cook for 5 minutes. Turn off the heat and cover the lid. Leave it to cook in the remaining heat for 15 minutes. Serve.

TIPS
1. Alternatively, in step 4, after cooking the pork belly over high heat for 10 minutes, turn to low heat and simmer for 30 minutes. Put in porcini mushrooms, dried cauliflowers and the soaking water. Cook over high heat uncovered to reduce the sauce. The sauce will be thicker and the seasoning will taste stronger this way.
2. Apart from pork belly, you may also use pork belly ribs for this recipe. They taste equally great.
3. Steam the pork belly before simmering, it will keep the shape of the whole pork belly and have a nice presentation.

Crispy chicken in black bean and olive honey glaze...120

INGREDIENTS

1 whole chicken (about 1.5 kg)
40 g preserved olives
8 g fermented black beans
120 g honey
caltrop starch

SEASONING

6 g salt
4 g sugar
4 g chicken bouillon powder
20 g caltrop starch
4 tbsp water (use 2 tbsp at one time)

METHOD

1. Rinse chicken. Remove the thigh bones, chest bone and backbone. Then chop into pieces. Set aside. Rinse the olives and black beans. Wipe dry and finely chop them.
2. Add 2 tbsp of water to the seasoning. Mix well. Stir into the chicken pieces and mix well. Leave them for 1 hour.
3. Coat the chicken in caltrop starch lightly. Heat some oil in a frying pan. Put the chicken in with the skin side down. Fry over medium heat until both sides golden and crispy. Set aside to drain excess oil.
4. Heat some oil in the same pan. Stir fry chopped olives and black beans until fragrant. Add 2 tbsp of water and stir well. Add honey and stir again. Cook until it bubbles. Turn off the heat and put in the chicken pieces. Toss to coat well. Serve.

TIPS

1. Remove the thigh bones, chest bone and backbones before chopping the chicken. It's more convenient to eat that way and it's less likely to choke on bones. The chicken will also be crispier.
2. Coat the chicken pieces in caltrop starch right before you fry them. If you coat them too early, the water drawn out of the meat will dilute and dissolve the caltrop starch so that the chicken won't have crispy crust after fried. I fry the skin side first because it helps draw the oil out of the skin. The chicken skin will also be crispier.
3. Turn off the heat and put in the chicken pieces to toss well when the glaze starts to bubble. This glaze is mostly honey without the thickening action of caltrop starch. First, honey shouldn't be cooked for too long. Besides, the moisture in the honey tends to make the chicken pieces soggy if they soak in the glaze for too long.
4. Apart from chicken you may also make pork belly ribs the same way. They taste equally great.

Crispy roast chicken stuffed with preserved clams and sand ginger...123

INGREDIENTS

1 chicken
40 g fresh sand ginger
35 g preserved clams
75 g ground pork
1 tsp light soy sauce
dark soy sauce
1/2 tsp caltrop starch
1 tsp sesame oil
1 tsp oyster sauce
1 tsp sugar
60 ml water

MARINADE

20 g coarse salt
20 g ground sand ginger

Sand ginger dip

80 g fresh sand ginger
25 g spring onion
120 g olive oil
2 g ground sand ginger
2 g chicken bouillon powder
2 g table salt
3 g white vinegar

METHOD

1. Rinse the sand ginger and spring onion. Drain well and let them air-dry.
2. Finely chop the sand ginger and spring onion. Mix all ingredients together. Serve as a dip with the roast chicken.

METHOD

1. Rinse the chicken and remove the innards. Wipe dry. Rub salt and ground sand ginger all over the chicken on both the inside and outside. Wrap in cling film and leave it in the fridge overnight.
2. Add light soy sauce and caltrop starch to ground pork. Mix well and leave it for 10 minutes.
3. Rinse the fresh sand ginger. Finely chop it.
4. Heat oil in a wok and put in preserved clams. Stir-fry until fragrant. Set aside. In the same wok, heat up a little oil. Stir-fry the sand ginger until fragrant. Set aside.
5. In the same wok, stir-fry the ground pork in some oil until done. Add preserved clams and sand ginger from step 4. Add water, sesame oil, oyster sauce and sugar. Mix well. Stir in caltrop starch slurry and cook until it thickens. Add a bit of dark soy sauce. This is the filling. Let cool and stuff the chicken with this filling. Seal the opening with a metal skewer. Wrap in cling film again and leave it in the fridge overnight.
6. Preheat an oven to 150°C. Roast the chicken for 60 minutes. Turn the oven up to 250°C and roast for 15 to 20 minutes more until golden and crispy. Serve.

Home-style braised beef shin in tomato sauce...126

INGREDIENTS

1.6 kg fresh beef shin
1 onion
4 potatoes (about 900 g)

TOMATO SAUCE

1 kg tomato puree
500 ml chicken stock
6 bay leaves
3 to 4 bird's eye chillies
6 tsp fish sauce
60 g sugar
1 tsp salt
4 tsp chicken bouillon powder
2 tbsp Tabasco
3 tbsp Worcestershire sauce
juice of 1 lime
ground black pepper

METHOD

1. Rinse the beef and cut into pieces. Blanch in boiling water. Drain and set aside. Peel and rinse the onion. Dice it. Rinse the potato. Peel and cut into pieces. Soak them in water to prevent discolouration.
2. Heat a pot and add oil. Stir-fry onion until fragrant. Put in beef shin and tomato sauce ingredients. Bring to the boil over high heat. Turn to low heat and simmer for 90 minutes. Drain the potatoes and add to the pot. Keep cooking over low heat for 30 minutes until the beef and the potatoes are tender. Serve.

TIPS

1. You can make this dish one day ahead. Just cover the lid and let the beef and potatoes pick up the remaining heat. Wait till the pot is cool and put it directly in a fridge. Reheat before serving the next day. It would tastes even more flavourful that way.
2. Generally speaking, potatoes should be cooked through and tender after being cooked for 30 minutes. Thus, they should go in the pot in the last 30 minutes. If you put them in too early, they'll melt in the sauce or break down into tiny bits. And of course that's not the intention.

Steamed pork ribs in Renren sauce...118

INGREDIENTS
600 g pork belly ribs (chop into pieces)
200 g homemade Renren sauce (see p.176 for method)
finely chopped spring onion

SEASONING
1 tsp sugar
1 tsp chicken bouillon powder
1/2 tsp salt
1 tsp oil

METHOD
1. Rinse the pork ribs and drain. Add seasoning and mix well. Leave them for 30 minutes. Add Renren sauce and mix again. Transfer onto steaming plate.
2. Boil water in a steamer. Put in the ribs and cover the lid. Steam over high heat for 8 minutes. Sprinkle with finely chopped spring onion. Serve.

TIPS
Marinate the pork ribs with seasoning for 30 minutes first. That would give the ribs background flavours upon which the Renren sauce builds on. Ribs made this way will be full of flavour and the unique taste of Renren sauce will be accentuated.

Double-steamed squab soup with black garlic, peanuts and black-eyed beans...158

INGREDIENTS
25 g black garlic (about half head)
60 g dried longans (shelled and pitted)
80 g peanuts
40 g black-eyed beans
1 squab (about 300 g)
1.8 litre water

SEASONING
15 g salt
6 g sugar

METHOD
1. Peel the black garlic. Rinse the dried longans, peanuts and black-eyed beans. Rinse and blanch the squab in boiling water. Drain.
2. Put all ingredients into a double-steaming pot or tall ceramic bowl. Cover with cling film. Boil water in a pot. Double-steam over low heat for 4 hours. Skim off the oil on the surface. Add seasoning according to your own taste. Serve.

TIPS
1. Make sure you cover the container with cling film to prevent condensation from falling back into the soup. It also retain the nutrients in the ingredients.
2. Most home-scale steamer or wok cannot hold enough water for 4 hours of continuous steaming. Thus, make sure you check the water level from time and time and refill it if necessary. Otherwise, it could be a fire hazard if it dries up.

Black garlic bread...130

INGREDIENTS

150 g black garlic (peeled)
150 g grated fresh garlic
1 1/2 tsp table salt
250 g Japanese butter
sliced baguette (or any other sliced bread of your choice)

METHOD

1. Mash the black garlic with a fork. Mix mashed black garlic, fresh garlic, salt and butter together until well incorporated. Transfer into a sealable container and seal well. Refrigerate until set.
2. Spread the black garlic spread on the bread. Preheat an oven to 180°C. Bake the bread for 3 to 4 minutes (depending on the type of bread you use and its thickness) until lightly browned and crispy. Serve.

TIPS

1. For this recipe, I prefer Japanese butter from brands such as Snow Brand or Meiji for their spreadable texture. It's easier to combine with other ingredients and the garlic spread tends to be creamier and tastier.
2. Before you grate or chop them, make sure you wipe the garlic cloves completely dry. Any uncooked water on them will make the black garlic spread go stale easily.
3. The ratio between black garlic and fresh garlic I put down here works well to my own taste and is just for reference only. Please follow this ratio once and taste it. Then adjust the ratio among black garlic, fresh garlic and butter according to your preference.

Black truffle shrimp toast...134

INGREDIENTS

4 slices sandwich bread
12 fresh shelled shrimps
180 g minced shrimps
15 g Tobiko caviar (Japanese flying fish roe)

SEASONING

salt
sugar
chicken bouillon powder
ground white pepper
caltrop starch

Black truffle sauce

1 frozen black truffle (about 30 g)
olive oil about 3 tbsp
salt

METHOD

1. Cut the crust off the sandwich bread. Then cut each slice into thirds. Set aside. In a bowl, mix together the minced shrimps, Tobiko and seasoning. Set aside. Butterfly each shrimp. Devein and sprinkle with a pinch of salt. Mix well.
2. Rinse the black truffle. Wipe dry and finely chop it. Add olive oil and salt. Mix well. This is the black truffle sauce.
3. On each piece of sandwich bread, spread on some minced shrimps. Put one shrimp over it. Heat oil in a wok and deep-fried the shrimp toast over medium heat until golden and crispy. Drain and leave them on paper towel to pick up excess oil. Arrange on a serving plate. Spoon a dab of black truffle sauce on top and garnish. Serve.

TIPS

1. To save time and effort, you may buy ready-made minced shrimps from the market. But it's not hard to make your own from scratch. Just devein the shrimps and crush them gently with the flat side of a knife. Finely chop them. Add seasoning and stir well. Slap the mixture onto a chopping board repeatedly until sticky.
2. For crispy deep-fried food, always turn the heat up to high in the last 10 seconds or so of the deep-frying process before draining. This last step helps draw more moisture out from the food so that it's crispier.
3. This shrimp toast is the perfect finger food for parties. You can make them 1 or 2 hours ahead of time and they'd stay crispy and delicious.
4. You can get frozen black truffle from shops specializing in mushrooms.

Kimchi and semi-dried oyster pizza...140

INGREDIENTS (Pizza crust)

300 g bread flour
20 g olive oil
1/2 tsp sugar
5 g salt
2.5 g instant yeast (dissolved in a little warm water)
200 g warm water

INGREDIENTS (Pizza toppings)

500 g cabbage Kimchi (squeezed dry, cut into pieces)
100 g tomato pasta sauce
200 g mozzarella cheese
6 to 8 semi-dried oysters (soaked in water till soft, rinsed, with the chewy round muscles on top removed, wiped dry and cut into pieces)
80 g Korean citron tea
Shaoxing wine

METHOD

1. In a mixing bowl, put in bread flour, olive oil, sugar and salt. Mix well. Add yeast and mix again. Slowly pour in the remaining warm water while kneading till the dough is formed. Then keep kneading until smooth. (If you're using a stand mixer, mix over low speed first until the dough is formed. Then turn to medium speed to knead till smooth.)

2. Put the dough into a container and cover with cling film. Let it rise at room temperature for 1 hour until the dough has expanded and becomes elastic. (If you make the dough in a dry season, spray some water on the dough after it is kneaded smooth. Cover it with a piece of damp paper towel to avoid drying out.)

3. Dust a counter with flour. Put the dough on top. Roll it out with a rolling pin into a round patty of the size and thickness that you prefer.

4. Dust a baking tray with flour. Put the pizza on it. Clean up the edges. Prick holes on the dough with a fork so that there won't be blisters after baked. Preheat an oven to 180°C. Bake for 10 minutes.

5. While the pizza crust is baking, heat up a wok. Put in Korean citron tea and Shaoxing wine. Mix well and cook until it bubbles. Put in the semi-dried oysters. Toss well and set aside.

6. Take the pizza crust out. Spread a layer of tomato pasta sauce on it. Top with a layer of cheese. Arrange Kimchi and semi-dried oyster on top. Sprinkle with cheese. Preheat the oven up to 220°C. Bake the pizza for 10 minutes until the cheese melts and the edge turns golden and crispy. Serve hot.

TIPS

1. Dissolve the yeast in water first before adding to dry ingredients. The yeast can bind with the dough better that way.

2. In step 1, do not add yeast and salt at the same time. You should mix the bread flour with olive oil, sugar and salt well before putting in the yeast. It's because the salt could kill the yeast so that the dough will not rise properly

3. Yeast is more active in a humid environment. Thus, do not put the dough in the fridge for rising. Make sure it rises at room temperature. You may have to adjust the rising time according to the temperature and humidity on a certain day.

4. Before you put the Kimchi on the pizza, make sure you squeeze it dry. Otherwise, the pizza will be drenched in its juice and won't be crispy.

Seafood angel hair pasta in Tom Yum sauce...137

INGREDIENTS

200 g angel hair pasta
6 live clams (rinsed)
6 medium prawns (rinsed, shelled and deveined) Save the head and shells for later use.
6 scallops (rinsed and wiped dry)
50 g fresh squid (rinsed and sliced)
5 Thai cherry tomatoes (cut into halves)
1/4 cup coconut milk

Tom Yum sauce

15 g shallot (coarsely chopped)
15 g taro (crushed)
2 stems lemongrass (use only the white part, cut into rings)
8 Kaffir lime leaves (rubbed and bruised with your hands)
70 g galangal (sliced)
2 bird's eye chillies (cut into rings)
60 g Tom Yum paste
600 ml water

SEASONING

freshly squeezed juice of 3 limes
25 g fish sauce
15 g sugar

METHOD

1. To make Tom Yum sauce, stir-fry shallot, garlic, prawn heads and prawn shells in a little oil until fragrant. Crush the prawn heads with a spatula to release the roe. Add water and all Tom Yum sauce ingredients. Cook for 30 minutes until the sauce reduces to 350 ml. Add seasoning and taste it. Strain the sauce and set aside.
2. Blanch the angel hair in boiling water for 3 minutes. Drain and set aside.
3. Blanch the clams, prawns, scallops and squid in boiling water or stir-fry them in some oil until half-cooked.
4. Boil the Tom Yum sauce from step 1 in a wok. Put in the angel hair to cook for 2 to 3 minutes. When the sauce starts to thicken, put in all seafood from step 3 and the tomatoes. Toss well. Add coconut milk and cook until all seafood is done and the sauce reduces. Serve.

TIPS

1. In step 3, the seafood should only be par-cooked till half-done. It's because it will be cooked again with the pasta later on. If it's cooked through in step 3, it will be way overcooked at last with rubbery and chewy texture.
2. In step 4, you may cook the sauce and the pasta according to your personal preference on doneness and consistency. You may add more or less liquid and cook it for longer or shorter as you desire.
3. When you add seasoning, you may also adjust the amount of condiments used according to your preference.

Egg noodles dressed in Renren sauce, garlic scallion oil and shrimp roe...144

INGREDIENTS

2 bundles egg noodles
1 tbsp garlic scallion oil
1 tbsp dried shrimp roe
2 to 3 tbsp homemade Renren sauce (see p.176 for method)
spring onion (finely chopped)

Garlic scallion oil

4 sprigs spring onion (cut into short lengths)
6 cloves garlic (crushed)
1/2 cup oil

METHOD

1. To make the garlic scallion oil, heat oil in a wok. Put in spring onion and garlic. Fry over low heat for 10 to 15 minutes until the spring onion and garlic turn golden and fragrant. Strain and discard the spring onion and garlic.
2. Boil the egg noodles in water until they loosen up and turn tender. Rinse in cold water. Reheat in boiling water again. Drain and save on a serving plate. Drizzle with the garlic scallion oil. Sprinkle with shrimp roe and top with Renren sauce. Sprinkle with spring onion at last. Serve. Toss and mix well before eating.

TIPS

1. Apart from egg noodles, you can actually dress any noodles or rice vermicelli of your choice the same way. They taste equally great.
2. I always rinse the egg noodles in cold water after boiling it. The thermal shock tends to make the noodles chewier and less sticky.
3. Reheat the Renren sauce in a microwave oven before using to soften the fruit pulp further. The sauce would blend in better with the noodles that way. The heat also melts the oil in it to make it more fragrant and delicious.

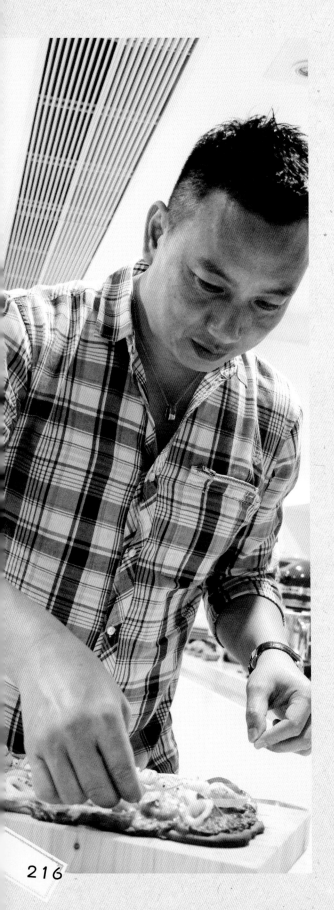

Tom Yum seafood pizza...150

INGREDIENTS (Pizza crust)
300 g bread flour
20 g olive oil
1/2 tsp sugar
5 g salt
2.5 g instant yeast (dissolved in some warm water)
200 g warm water

INGREDIENTS (Pizza toppings)
150 g squid (cut into rings or strips)
10 medium prawns (shelled, deveined)
4 Hokkaido giant scallops (cut each into quarters)
Tom Yum paste (amount depends on your tolerance to spicy food)
100 g tomato pasta sauce
200 g mozzarella cheese
1 sprig Thai basil (leaves only)

SEASONING
salt

METHOD

1. To make the pizza dough, mix bread flour, olive oil, sugar and salt together. Add yeast and mix further. Pour in a little warm water at a time while kneading to form dough. Then knead until smooth. (To use a stand mixer, attach the dough hook. Mix with low speed first until the dough is formed. Then turn to medium speed and knead till smooth.)

2. Put the dough into a bowl. Cover with cling film. Let it rise at room temperature for 1 hour until the dough has expanded and turn elastic. (If you make the dough in a dry season, spray some water on the dough after it is kneaded smooth. Cover it with a piece of damp paper towel to avoid drying out.)

3. Dust a counter with flour. Put the dough on top. Roll it out with a rolling pin into a round patty of the size and thickness that you prefer.

4. Dust a baking tray with flour. Put the pizza on it. Clean up the edges. Prick holes on the dough with a fork so that there won't be blisters after baked. Preheat an oven to 180°C. Bake for 10 minutes.

5. While the pizza crust is baking, wipe the squid, prawns and scallops dry. Add a pinch of salt and mix well. Fry them in a pan with a little oil until half-cooked. Drain and set aside.

6. Take the pizza crust out. Spread a layer of Tom Yum paste and tomato sauce on it. Arrange squid, prawns and scallops evenly over. Sprinkled with grated mozzarella cheese at last. Preheat the oven up to 220°C. Bake the pizza for 10 minutes until the cheese melts and the edge turns golden and crispy. Remove from oven. Sprinkle with Thai basil leaves. Serve hot.

TIPS

1. Dissolve the yeast in water first before adding to dry ingredients. The yeast can bind with the dough better that way.

2. In step 1, do not add yeast and salt at the same time. You should mix the bread flour with olive oil, sugar and salt well before putting in the yeast. It's because the salt could kill the yeast so that the dough will not rise properly

3. Yeast is more active in a humid environment. Thus, do not put the dough in the fridge for rising. Make sure it rises at room temperature. You may have to adjust the rising time according to the temperature and humidity on a certain day.

4. I marinate and stir-fry the seafood until half-cooked before putting them on the pizza. Of course, par-cooking them helps shorten the baking time of the pizza. Another reason for doing so is to cook some of the moisture out of the seafood so that the pizza won't be drenched in their juices.

5. Tom Yum paste is the key condiment used in Tom Yum Goong soup. Apart from making soup, you can also marinate meat and season any dish with it (such as fried rice). You can get Tom Yum paste from any Thai grocery store.

Clams in spicy wine sauce with instant noodles...147

INGREDIENTS

900 g clams
1 pack instant noodles (any brand you prefer)
30 g shallot (finely chopped)
30 g grated garlic
5 bird's eye chillies (cut into rings)
3 to 4 sprigs coriander (rinsed, cut into short lengths, some finely chopped)
1 cup chicken stock
25 ml Chinese rose wine
15 ml Shaoxing wine
15 ml rice wine
80 ml oil

SEASONING

50 g satay sauce
1/3 tsp salt
1/2 tsp sugar

METHOD

1. Soak the clams in salted water for 2 to 3 hours until they spit out sand. Scrub them well with a brush. Drain and set aside.
2. Heat oil in a wok. Stir-fry shallot until fragrant. Add garlic and bird's eye chillies. Stir until fragrant and the garlic is browned lightly. Add satay sauce and mix well. Put in the clams and stir further. Add chicken stock, Chinese rose wine, Shaoxing wine and rice wine. Bring to the boil again. Add salt and sugar. Cook until all clams pop open. Add coriander and toss well. In the meantime, boil another pot of water and put in the instant noodles. Cook until done. Strain and put the noodles into a serving bowl. Pour the clams and the spicy wine sauce over the bed of noodles. Sprinkle with chopped coriander. Serve.

TIPS

1. Apart from clams, you may cook other seafood such as mantis shrimps, conches, prawns and crabs the same way. If you're not a fan of instant noodles, feel free to use other starch-rich staples, such as mung bean vermicelli, udon, kuzukiri (Japanese kudzu noodles), rice noodles or even deep-fried dough sticks instead. As long as the staples suck up the spicy wine sauce, they'd work just fine.
2. I've tried making a hotpot soup base with this recipe by adding lots of radish, shiitake mushrooms, Chinese celery and tofu. It was divine. Feel free to try on the cold winter days.
3. You may use more or fewer bird's eye chillies according to your personal preference.

Black truffle soup with mushrooms...160

INGREDIENTS

80 g frozen black truffles
150 ml whipping cream
150 ml milk
150 ml chicken stock
10 g butter
grated garlic
Japanese beech mushrooms
parsley (finely chopped, as garnish)

ROUX

20 g butter
20 g flour

SEASONING

salt
sugar
chicken bouillon powder
truffle oil

METHOD

1. To make the roux, heat a pot over low heat. Put in 20 g of butter and heat until it melts. Stir in the flour into a thick paste. Set aside.
2. Thaw the black truffles and rinse briefly. Wipe dry with paper towel. Slice part of it and set aside. Finely chop the rest. Put chopped black truffle into a blender. Add chicken stock and blend until fine. Set aside.
3. Pour whipping cream and milk into a pot. Bring to the boil. Then add the truffle chicken stock from step 2. Bring to the boil over low heat. Season with salt, sugar and chicken bouillon powder. Taste it and season accordingly if needed. Stir in the roux from step 1 and keep stirring while heating it up. When your desired consistency is achieved, pour into a soup tureen or serving bowls.
4. Melt 10 g of butter in a pan. Stir-fry grated garlic until fragrant. Add beech mushrooms and stir until tender and cooked. Arrange over the soup. Top with sliced black truffle and drizzle with some truffle oil. Sprinkle with parsley. Serve hot.

TIPS

1. Apart from beech mushrooms, you may also use any other mushrooms as you wish.
2. When you blend the black truffle with the chicken stock, blend as finely as you can. Otherwise, the black truffle will be too lumpy and fibrous. The soup won't end up as creamy and velvety as it should be.
3. You can get frozen black truffle from shops specializing in mushrooms. Adding a dash of truffle oil at last helps heighten the aroma. You can get truffle oil from large-scale Japanese supermarkets. You can make 4 to 6 servings of soup with this recipe.

Tom Yum Goong pumpkin seafood soup...154

INGREDIENTS

10 medium prawns (shelled, deveined, heads and shells set aside for later use)
1 mud crab (dressed, chopped into pieces)
10 clams (rinsed, soaked in salted water)
5 cloves garlic (sliced)
5 shallots (finely chopped)
8 stems lemongrass (use only the white part, finely chopped)
40 g galangal (sliced)
6 sprigs coriander (finely chopped, set the roots aside for later use)
2 to 4 bird's eye chillies (the amount depends on your tolerance of spicy food, finely chopped)
10 g Kaffir lime leaves (rinsed, rubbed with your hands)
450 g pumpkin (peeled, de-seeded, cut into chunks)
250 g Thai cherry tomatoes (rinsed, cut in halves)
10 straw mushrooms (rinsed, cut in halves)
2 litres chicken stock

SEASONING

150 g Tom Yum paste
20 g fish sauce
10 g sugar
freshly squeezed juice of 4 limes
coconut milk (added last)

METHOD

1. Heat 1/2 cup of oil in a pot. Stir fry garlic and shallot until fragrant. Put in the prawn shells and heads. Fry until fragrant. Crush the prawn heads with a spatula to release the roe. Add chicken stock, lemongrass, galangal, coriander roots, bird's eye chillies and Kaffir lime leaves. Bring to the boil over high heat. Cover the lid and turn to low heat. Simmer for 30 minutes. This is the soup base.

2. Strain the soup base. Pour it back into a pot. Add pumpkin and cook for 10 minutes until tender. Add tomatoes, straw mushrooms and seasoning. Mix well and taste it. Season further if necessary. Cook until the pumpkin is mushy and the tomatoes are soft. Put in crabs and clams. Cook further. Add prawns and cook until done. Pour in some coconut milk and sprinkle with finely chopped coriander. Stir and serve hot.

TIPS

1. Tom Yum paste is a readymade condiment available from Thai grocery stores. It's a mixture of lemongrass, galangal, chillies, peanuts, dried shrimps, garlic, shallot and tamarind juice that have been blended and fried in oil. There are many brands of Tom Yum paste, each of which differs in sourness, saltiness, sweetness and spiciness. If you're making Tom Yum dish the first time, you may buy different brands and taste it to find your favourite one.

2. Tom Yum Goong soup is a Thai classic and the red oil from fried prawn roe that floats on the soup is the key attribute that tells an authentic Tom Yum Goong soup from a half-hearted imitation. Thus, the best prawns to be used in Tom Yum Goong are the giant river prawns with huge heads as they have the most roe in their heads. Yet, it's not easy to find giant river prawns. So I settled for medium marine prawns. For the best result, make sure you fry the prawn heads long enough in step 1. Try to release the roe to give the soup flavour, aroma and zest.

3. You may use the Tom Yum soup base for hotpot. Any seafood, meat or noodles would taste great in it.

Warm black truffle chocolate fondant a la mode...163

INGREDIENTS

80 g unsweetened dark chocolate, 80 g milk chocolate, 90 g unsalted butter, 50 g sugar, 4 eggs, 4 egg yolks, 60 g cake flour, 20 g black truffle, 20 g truffle oil, milk or vanilla ice cream

METHOD

1. Put dark chocolate, milk chocolate and butter into a bowl. Put the bowl over a pot of simmering water. Keep stirring until chocolate melts. Set aside. Sieve the flour into another bowl. Finely chop the black truffle. Set aside.
2. Beat sugar, eggs and egg yolks until well blended. (Just beat until they mix well. Do not over-beat it till stiff or beat too much air into the mixture. Otherwise, the cake will expand and puff up after baked, and this is not desirable for this recipe.)
3. Pour the chocolate and butter mixture into the egg mixture. Beat again. Sieve in the flour. Add black truffle and truffle oil. Fold gently until lump-free. Pour into ramekin or cake tin up to 70% or 80% full.
4. Preheat an oven to 200°C. Bake the cake for 10 minutes. Check doneness by pressing the edge of the cake with your finger to make sure it has set. Then insert a bamboo skewer into the cake about 1.5 cm away from the cake tin. If the skewer comes out with liquid batter, the cake is done.
5. Turn the cake out of the ramekin or cake tin. Put a scoop of ice cream on top and garnish. Serve.

TIPS

1. Try to chop the black truffle as finely as possible. The texture of the cake will be finer and smoother that way.
2. Do not use bread flour for this recipe. Otherwise, the cake will be too tough and chewy after baked.

Okinawa Kurozatou New Year Cake...166

INGREDIENTS

500 g Okinawa Kurozatou (dark brown sugar) 1 litre water, 830 g glutinous rice flour, 15 g oil

METHOD

1. Boil Kurozatou in water until it dissolves. Pass it through a mesh sieve to remove any impurity. Let cool.
2. Sieve the glutinous rice flour into a big mixing bowl. Add Kurozatou syrup little by little into the glutinous rice flour while stirring until a smooth batter is formed. Add oil. Keep on stirring until the oil is well-incorporated. Pass the batter through a mesh sieve to remove any lump.
3. Lightly grease a steaming tray or tin. Pour in the batter and cover with cling film. Boil water in a steamer. Steam the cake for 90 to 100 minutes. Let cool and refrigerate for 2 days so that the cake firms up. Slice and fry in a pan over low heat until it softens and is heated through. Serve.

TIPS

1. When you mix the Kurozatou syrup with glutinous rice flour, make sure you stir the mixture until well combined. Otherwise, the dry lumps in the batter will make the cake patchy and rough in texture.
2. Adding oil to the batter makes the cake more velvety in texture and shinier in appearance.
3. The depth of your steaming tray and the heat of your stove may vary. Thus, you may have to adjust to steaming time according to the situation.
4. To check the New Year cake for doneness, insert a chopstick into the centre of it. If the chopstick comes out with some pale batter sticking on, that means the cake is not done yet.
5. Greasing the steaming tray before pouring in batter prevents the cake from sticking to the tray so that it's easier to turn out after steamed.

Homemade Chinese plum wine...170

INGREDIENTS
4.5 kg fresh Chinese plums
3.6 kg rock sugar
1 wide-mouthed airtight glass jar

METHOD
1. Rinse the plums. Drain and wipe them completely dry.
2. Make a few crisscross light incisions on the plums.
3. Pound the sugar in mortar and pestle. Put a layer of rock sugar on the bottom of the jar. Top with a layer of plums. Repeat with alternating layers of sugar and plums until the jar is filled.
4. Leave the jar covered at room temperature. The juice that seeps out from the plums will mix in with the rock sugar and turn into a syrup. Leave it to ferment slowly.
5. A week later, there should be no rock sugar left. The liquid turns from clear into dark ruby red. There will be bubbles rising from the liquid, meaning active fermentation has begun. The fermentation may take several months and the sugar is the food that feeds the wild yeasts which turn the syrup into alcohol over time.
6. After 10 months or 1 year, the fermentation will be complete and the mixture of plum juice and rock sugar will become plum wine with exquisite aroma and flavours.

TIPS
Make sure you wipe the plums completely dry before using. The glass jar should also be sterilized in boiling water and then dried thoroughly. Any water in the utensils or the plums would lead to the growth of harmful bacteria and moulds.

傳說的國度

傳說中，在浩瀚宇宙的某個空間，存在着一個貓之國度，為了逃避豺狼星虐貓族的滅絕追殺，他們將國度秘密遷徙到地球，然後隱匿在東京的一處角落，而只有屬於喵星人國度的子民，才能知道他們現時藏身的所在地。

我從香港分會部落取得秘密資料，這個充滿神秘的貓之國度，就在東京。台東區谷中三丁目，只要你從新宿站的三號的神秘月台，乘搭JR喵星號列車，在穿越銀河之後的日暮里站下車，沿着站裏西面出口的隱形地圖指示（當然是只有喵星國的子民才會看到），最重要就是留意頭髮背後長有貓耳朵的人，只要跟着他們，就可以來到這個喵星人的世界國度，去尋找你想要的，有關貓咪的所有東東！喵。
=^_^=

童心

我一向都有觀察別人的習慣。

我曾看見地鐵上一位持着拐杖、步行有問題的殘障人士，面對攘熙人群感到慌措不安，但他最後鼓起了勇氣，邁出腳步接受眾人帶着奇異的眼光。

我也曾發現一個咬着香煙，身上佈滿紋身，古惑仔造型的青年，有着一顆柔軟的心，當他給流浪貓狗餵食的時候。

我也有看光的習慣。

我偏愛暖黃的光，十五歲的一個夜晚，我曾看到過最憂傷的光。它只是一盞昏黃的路燈，閃爍着像是隨時都有可能熄滅的光，它昏暗地照在寂寞的街道上、布滿塵埃落葉、久無坐客的椅子上，和我仍然帶着稚氣的臉頰上。

我想，也許明天我都不會出現在這個地方了，我正在探索着我還未知的、充滿希冀的將來。而它卻要在這被人遺忘的角落裏散發出破舊的燈光，企圖照亮這被遺忘的一切。

於是十五歲的我心想，我要記得它們，在這樣一條街道上有這樣的光散發過昏黃的光芒。它的存在便得到了完整。

我不是多愁善感。

我只是有點敏感，我只是很勇敢的道出大多數人在成年以後便缺失的一部分，這一部分因為被人遺缺而顯很珍貴的一種天真。而大多數人領悟到的只有這種誤以為是多愁善感的怪異。

我希望某天，當我走在街上的時候，也有與我一樣的人會看看我，他們肯定猜不到，我是這樣一個一把年紀仍心存童心的人。

這裏有同路人嗎？

咖啡，
如果沒有你，日子怎麼過？！

在咖啡國度裏，咖啡，可以說是一種合法的興奮劑，她擁有一種無形的魔力，讓這國度的子民，屈服於她濃厚的汁液和瀰漫滿室的香氣之中。

當三五知己，在一間喜愛的咖啡室裏，彼此交換着或深或淺的心事，又或者是那些無無聊聊的生活鎖碎，咖啡也是就很多人生活的一部份和寄託。

總有些時候。你不想多說話，不想做任何事情，也不需要其他人陪伴，就只想一個人在咖啡店裏呆着，享受一杯咖啡的美妙時光，感受一下平日匆匆忙忙而遺忘了的生活氣息。

九份 。 遊記

和去台北的路相反，計程車往山頂的方向駛去。一路經過了許多山和樹、一座氣魄恢宏的廟宇和零星點點的建築物，到達了九份。

九份於我最早的印象，是多年前侯孝賢的電影「悲情城市」，因為電影效應的關係，很自然地，這個主要場景，瞬間便變成了到台北必到旅遊景點。

九份我已來過很多次了，06年我主持過的一個飲食節目，也在這裏拍攝過，亦留下了很多回憶片段。轉眼那麼多年過去了，每次來台北，我都盡量抽時間到來逛逛。

雖然現在多了更多遊客，但九份那種老台灣的味道還是非常吸引着我。雖然正值春季，山間雨霧環繞，雨下的九份，顯得有點淒惶。

車停在九份牌坊旁，踏上這片土地時，涼意從頭到腳伸開，我又來到這裏了，我喜歡這個地方。

我是來收集記憶的。

古老蒼舊，迂迴曲折的街道，佈滿劃痕的牆壁和石板路上，飄着各種台式美食的香氣，叫賣聲、遊人的喧雜聲混成一片。一隻缺失了一條腿的狗，店舖老闆娘熱情招待狗狗「寶貝，進來歇一會。」一切都自然而然，絕無虛情假意。

雨正變小，逛着逛着，來到一間共有兩層，看得見整個九份風景的咖啡店坐了下來。群山在餐桌旁展開，遠處透雲透霧的陽光零零散散爬上山巔，天開始轉晴了。

雲霧開始減少，動人心魄的景色慢慢顯露出來。山與山之間的薄霧像是一層神秘面紗，被風輕輕推着緩慢離開。我睜大眼睛，把這幅精美的動態畫作，刻錄在我的腦海和攝錄在我的手機裏。

我走在電影的場景，每踏上一塊石階，便像經歷了這裏曾發生過的遙遠悠長的故事，或悲傷或淒美或讓人欣然一笑，又或是那些家常瑣碎，一切都那樣美好並持續地上演着。

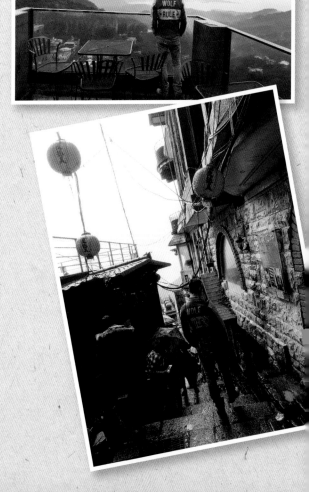

或許，日後我還將去到更遙遠的地方，看到新的景物，聽到不一樣的故事。但於此發生的一切都將成為屬於我的美好的回憶。

和來時的路相反，計程車往山底的方向駛去。我兩手空空，却感覺滿載而歸，又再經過那座氣魄恢宏的廟宇，經過許多的山和樹，我離開了九份。

異類

我在冬日的路上走着，寒風凜冽，一團白雲迎面向我飄來，在距離地面幾尺的高度，像低飛的幽浮，在我的面前緩慢停了下來，圍繞着我。

「先生，可否給我喝點水」？

白雲輕聲細語地說：「我實在很口渴」。

白雲隨着我來到便利店，我買了一瓶礦泉水，付錢的時候，店員望着白雲說：「這朵雲真漂亮啊」！

我撐開瓶蓋，看着礦泉水從瓶裏上升起來，慢慢蒸發在空中。

我說：「白雲，妳叫甚麼名字呀？妳怎麼會獨自來到地上呢」？

白雲有些遲疑，雲裏夾雜着的烏雲在微微顫動。

「我叫小白，它們說我是異類，因為我不願喝海港的污水，不願認空中的霧霾為親友。於是它們便把我分離開來，我飄蕩在地上，我知道這樣會很危險，可是我實在太口渴了」。

小白說完繼續吸允着瓶中的礦泉水，我看着實在覺得這朵雲很可愛，正要伸手撫摸它的時候。。

這時一陣兇惡的猛風突然吹來，夾帶着霧霾，將小白撲散在空氣中。瞬間遠去的惡風回頭狠狠地看了我一眼，並發出凶惡呼嘯：「異類！異類」！

我們是異類嗎？正當我這樣思索着，一個聲音從四處聚集過來。

「先生。先生。是我！」

「是小白嗎？你在哪裏？」

「我在任何地方。只不過我不再是一朵雲，我已經被蒸發成水蒸氣。先生，

我很快就要消失了，謝謝你！」

「小白，你聽我說，你並不是異類！先生生存在這樣的世界裏，深有體會。有時候我們並沒有作出選擇，我們只是做回一個真實的自己！」

我伸出了手，「小白，與你有了這樣的遭遇，也許我也變成了某些人眼裏的異類。但這不足以成為放棄自我的藉口，至少我的心是愉悅的。我現在撫摸到你了嗎？」

「先生，我就在你的手心裏。我現在已經很微弱，請你帶我離開好嗎？謝謝你…」

我在路上走着，寒風凜冽。

攥緊手心，握着堅定的自己，活在這樣的世界裏。

零失敗的煮人

作者	Author
余健志	Jacky Yu
策劃/編輯	Project Editor
	Catherine Tam
攝影	Photographer
	Johnny Han
	Jacky Yu
	Amelia Loh
美術統籌及設計	Art Direction & Design
	Amelia Loh

出版者　Publisher

Forms Kitchen

香港鰂魚涌英皇道1065號
東達中心1305室

Room 1305, Eastern Centre, 1065 King's Road,
Quarry Bay, Hong Kong.

電話　Tel: 2564 7511
傳真　Fax: 2565 5539
電郵　Email: info@wanlibk.com
網址　Web Site: http://www.formspub.com
　　　　　　　http://www.facebook.com/formspub

發行者　Distributor

香港聯合書刊物流有限公司

SUP Publishing Logistics (HK) Ltd.

香港新界大埔汀麗路36號
中華商務印刷大廈3字樓

3/F., C&C Building, 36 Ting Lai Road,
Tai Po, N.T., Hong Kong

電話　Tel:　2150 2100
傳真　Fax:　2407 3062
電郵　Email: info@suplogistics.com.hk

承印者　Printer

百樂門印刷有限公司

Paramount Printing Company Limited

出版日期　Publishing Date
二〇一六年七月第一次印刷　First print in July 2016